PASSAGEWAYS

PASSAGEWAYS

✦

Dialogues from Beyond the Rubicon, Volume 1

Russell J. Cordua, Jr.

iUniverse, Inc.
New York Lincoln Shanghai

PASSAGEWAYS
Dialogues from Beyond the Rubicon, Volume 1

iUniverse books may be ordered through booksellers or by contacting:

iUniverse
2021 Pine Lake Road, Suite 100
Lincoln, NE 68512
www.iuniverse.com
1-800-Authors (1-800-288-4677)

Because of the dynamic nature of the Internet, any Web addresses or links contained in this book may have changed since publication and may no longer be valid.

The views expressed in this work are solely those of the author and do not necessarily reflect the views of the publisher, and the publisher hereby disclaims any responsibility for them.

ISBN: 978-0-595-46313-8

Printed in the United States of America

To my mother, and to all the children, those here and those to come.

"When we tug on a single thing in Nature, we find it attached to everything else."

—John Muir

Contents

Prologue

(Southern France—approximately 30,000 years before the birth of Christ ... the time of Cro-Magnon Man)

My brother and I are the crippled, the injured, the weak of our now disbanded group. As the chilly dawn creeps forward, the two of us huddle together deep within the inner walls of this unfamiliar cave, a faint flicker from a small cooking fire burning just beyond the mouth of our cavernous shelter. The dim glow from an oil lamp illuminates the remainder of the cave behind us. We are the unfortunate casualties from the most recent and most brutal search for new hunting grounds. Our bodies, more than the other warriors, have been, by sheer chance alone, beaten by our travels just a bit more severely. The others are still effective, still useful out upon the open terrain; we're not, and so we have been left behind to wait for an improbable return. And in these long stretches of silence, each of us realizes all too well our most likely fate. In what was surely a final show of compassion, the others had left for us the last significant kill—a third of a Bison carcass which, at this point, would provide maybe one substantial meal.

I stare down at my swollen leg, the blood caked in heavy, smeared black hair now matted and rust colored, and notice the greenish hue slowly gathering just below my skin. Closing my eyes, I feel a dull sensation of throbbing, but thankfully not much more.

I look over at my older brother of three years. We had both been severely injured during the last battle, a battle which had taken place just three days before. As the warriors of our tribe, about fifteen in all, we were in a desperate trek westward in search of more fertile hunting grounds when we found ourselves face to face with another tribe who themselves were amidst a similar search. The fighting began without warning. A group of about ten warriors from this unfamiliar tribe attacked the lead eight of us as we moved through the thick forest. What they didn't know was that there were another seven of us trailing just a few moments behind. Unaware that they would be quickly outnumbered, they jumped us from out of the underbrush, instantly killing two in our group. The other six of us, which included my brother and I, managed to fight them off until the others quickly appeared. What followed was an absolute slaughter. All eight

1

in their pack were killed in a matter of minutes, along with one more from our group. My brother and I were both badly injured. We both had been impaled with spears made from reindeer antlers—he in the calf, me in the thigh. Neither of us could walk on our own. And this, we both knew, meant certain death.

The others carried us the rest of the way through the forest, heading for the small range of mountains we could see rising in the distance. We all knew that there would be shelter there, but the question remained: would we have to fight another tribe already dwelling there? Or would the shelter already be abandoned? We had left our women and children in our own caves, leaving our tribe's other dozen warriors with them for protection, as we struck out to find a new home in a more abundant area. Had whatever tribe that inhabited the caves and rock shelters in the mountainous range before us done the same? That was the danger we faced as we began to ascend the rocky mountainside.

Though shaking with fever, I can still remember my brother and I being set down behind a rocky ledge at the foot of the mountain, hidden away. The others, in cautious silence, proceeded to scale the mountain, their weapons at the ready. As my brother drifted off, his eyes closed, I watched as the others climbed their way to an upper ledge and then, after motioning silent commands, they scattered in different directions. In the next moment, I heard the wild screams of battle pouring down from the sky above, echoing off the rocks, penetrating my raw, frightened nerves, and I trembled.

After a few moments had passed, the leader of our pack suddenly appeared on the ledge above, motioning for the others to retrieve us. At that point, all I can remember is falling into a fevered sleep as I was hoisted upon a shoulder. Some time later—how much later I have no idea—I awoke to find myself inside this cave with all the other warriors, the fresh smell of death filling the air. The cave had obviously been recently occupied; the floor was covered with animal skins, and there was an impressive collection of tools, such as scrapers, cutters, pointed burins and sewing needles—some made of bone, some of wood, but most made of antler. Located at the deepest part of the cave was a small tent, illuminated by what looked to be a small oil lamp. The former occupants, five in all—three men and two women—had been removed and killed by our warriors just a few moments before; this I could tell because the scent of them lingered.

The rest of our group stayed with us for the remainder of the day and night. Five of them went out on the hunt and, to everyone's joy and amazement were able to secure a Bison carcass after only a few hours. A kill of that magnitude had become relatively rare, so that evening the group enjoyed what seemed like an incredible feast.

After a night of unrelenting pain and fever, my brother and I awoke to find ourselves alone. At the break of dawn, the rest of the group had continued on in the search for new homelands. We knew the chances of them returning for us was improbable at best—even if they were to find new homelands, probably only enough of them would survive the journey to be able to return to retrieve the women and children. We would be an afterthought at best. Our only hope was for them to find a potential homestead nearby, so that returning for us would be easy and quick for them to do. Every day that passed without their return, however, would indicate that their search was driving them further and further away, and thus sealing our fate.

That first day, we both lay still in our suffering, drinking sparingly from the wooden container of water left for us and nibbling even more sparingly at the carcass. My brother was in worse shape than I—the skin of his calf was badly torn, ripped down, in fact, to the cartilage. One could see the exposed bone protruding through the wound if he moved in a certain way. The only thing we had going for us was the cool weather; the hot summer temperatures had recently ended, leaving a sustained period of comfortable temperatures in its wake. The nights, in fact, were turning cool, but not unbearably so. Our group, we knew, would be able to travel quickly in this kind of weather.

The first night alone was nothing more than a hot flash of pain and sweat that seemed to never end. Both of us were in such intense discomfort that we slipped in and out of powerful delusions. My brother wailed out in a hallucinatory, violent sleep throughout most of the night until he finally shook and shivered into a kind of motionless shock. I, on the other hand, trembled with flashes of chaotic dreams and visions until I finally fell into a black, frenetic sleep. My wound was wide and gaping, and the pain felt like hot needles searing open nerves. I dreamed about fire on my skin.

The following morning, I finally forced myself to my feet and hobbled about the cave, eventually making my way back to the tent. It was a relatively small structure made of bison and reindeer hides. I picked up the burning oil lamp that sat just outside it, pulled back the flap, peered in at a small gathering of tools—mostly scrapers, perforators and knives, most of which were made from ivory. There was also a burin, a stone bowl filled with grease, apparently for the oil lamp, two empty skulls, three spears and, most significantly, three bowl-like containers made from human skulls. One bowl contained ocher (an earthy clay), the other charcoal, and the third manganese oxide. Sitting beside these bowls were engraving tools and several paint brushes, the tips of which were made from animal hair. Struggling in growing pain and exhaustion, but suddenly excited, I scooped up the paint brushes and the engraving tools, hobbled back to where my brother lay, still unmoving and still fading in and out of consciousness. I grunted loudly and expansively as I approached him, holding what I had found out in front of me. My brother was a master artist and was revered by many in the tribe as a shaman, in large part because of the powerful images he often created.

My brother looked up, his gray and cloudy gaze suddenly narrowed when he realized what I held in my hands. He let out a deep groan, shifted his body to get a better look. I nodded my head up and down in confirmation as he studied the implements. I quickly set them down before him, motioned toward the back with a loud grunt of my own and headed off to retrieve the other items. When I returned, my brother was standing, balanced on one foot, an engraving tool dangling in his fingers while his free hand gently ran across the surface of the cave wall, studying it. Having been his apprentice on many occasions before, I quickly began my work. First, I took a stone hammer and poured a portion of the ocher onto a flat rock and began to pound it until it was reduced to a fine powder. Then I did the same with the charcoal and the manganese oxide. At this point, when all were ground sufficiently, I poured each into separate bowls, stood and urinated equally into each of them. From there, I took a knife and mixed the ingredients of each bowl until what remained were three different colored pigments. Once I felt I had stirred them to the proper texture, I placed the bowls on the ground before my brother, kneeled down beside him and held up the oil lamp so the portion of the wall he was examining would be illuminated. From then on, for however long he worked, I would move as he needed me to move.

Without first outlining, as he normally would do, he moved right into his work by beginning as high up on the wall as he could reach, and painted a large sun. Then he immediately turned his attention to the very bottom of the cave wall, painting a piece of earth with sprouting grass and bushes. He worked diligently, taking short breaks upon completing each section of this particular scene. He would nibble at the quickly dwindling bison carcass and sip some water before pulling himself to his feet to start again. After one longer than usual break, shivering all the while with fever, he returned to the painting of the earth and drew a worm pushing its way from out of the soil. From there, he painted a small bird flying in a sharp, downward descent. Then he moved a few feet to his right to a clean section of the wall, and repainted the sun. Then he repainted the soil and the grass and the bushes, finishing with the bird ascending upward, the worm dangling from its tiny beak. At that point, with many hours having passed, he finally slumped down in an exhausted heap, and I knew he was finished for the night.

◆   ◆   ◆

As I stir awake, my brother and I huddled together, I can feel his heavy breathing as he sleeps. As the pain penetrates my every move, I slip out from under him, move to the disintegrating bowl of water, splash a tiny bit on my face before taking one long drink. Then I carry the bowl over to my brother and sprinkle a little water on his face and forehead to cool his sweltering skin. When I'm finished, I find myself studying the sleeping man in the dim light of the oil lamp. My brother has always been an extraordinarily strong man with dark features, from his square, scraggly bearded jaw line, to his black eyes, to his creased forehead, to his perennially intense scowl. Our father had been a fierce hunter and warrior who had been killed some years back during a battle with a rival tribe. Though our father was one of the most revered of all the hunters, my brother was the most revered of the artists. He and his drawings would often be at the center of not only initiation ceremonies for the young males, but tribal rituals to honor the spirits of ancestors long passed. On these occasions, my brother, wearing the skin and head of a bison, would often play a flute made of carved bone as we chanted, prayed, and danced at the foot of his drawings.

As my brother continues to sleep, I gather as much of my strength as I can, and begin to hop out toward the cave's entrance. It's a relatively deep underground structure, but I manage to make it up to the mouth of it within a few minutes. As I warily peek out I notice that the early morning sun, though still low upon the horizon, is blazing, and I am momentarily blinded by it. After my eyes adjust, I find myself still cautious in my advancement, mostly because I don't want to be spotted by another tribe that may be passing by. This is the danger for me at the moment. But after surveying the area, from the forest immediately below me to the open stretches of terrain well into the distance, I see that all is clear. Looking around one last time, I venture out very slowly. The air is still and pleasant and I find myself gulping it in like fresh, clear water, and it feels so good that I am suddenly terrified by my inevitable demise. I realize all to well that this could very well be the last time I am to feel the warmth of the sun, taste the purity of the air, see the horizon.

Pushing these thoughts away, while at the same time being motivated by them, I gaze around to see if there is any source of food or water that I could have possible access to in my hobbled condition. But there is nothing, only a rocky ledge that bends around the mountainside possessing only this one cave. I look at my torn leg, disheartened with the knowledge that food and water is only a relatively easy climb, though long and steep, down into the waiting forest. But in my weakened condition I would surely tumble to my death. Our only hope, my journey to the outside has now reaffirmed, is for the others to return for us.

Limping back to the security of the cave's inner sanctum, I find my brother on his feet, brush in hand, painting another sun next to the last scene. He paints rapidly, as his strength is beginning to fail him. I quickly feed the lamp with grease to keep it burning, move in position, hoist the lamp up for the proper lighting, and silently watch as he paints the same bird as before, but this time perched on what looks like a low tree limb, the worm, half eaten, still dangling from its beak. Once this is completed, he proceeds to paint, high up near the sun, a larger bird, an osprey, on a downward descent of its own. When this is finished, he again moves and repaints the same scene, but this time with the osprey in upward flight, the smaller bird in its clutches.

The next scene is again of the osprey, still under the same sun, having returned to its nest on a high tree limb, feeding its young from its own beak. I watch my brother as he holds his weak, trembling body as still as he can while his eyes narrow in concentration. He moves on, painting the same sun, as well as the same tree limb holding the full osprey nest, but this time the osprey is gone, probably out on another hunt. Once this scene is complete, I watch in fascination as he paints darkened storm clouds with a bolt of lightning penetrating through them, a bolt of lightning that extends down onto the trunk of the tree. In a fevered pitch, breathing heavy, but willing himself to go on, he begins another painting, the hours slowly passing. He paints the sun and the same tree, but the trunk of the tree is split, and the bird's nest is now lying upon the ground with three baby birds fluttering about helplessly. He wipes the sweat from his brow, continues. Off behind the trunk of the tree, he paints a small fox, crouched low and ready to spring. The next scene, under the same sun, is of the fox devouring the tiny birds, unaware that just behind him on the ground is a man-made snare.

With this painting complete, my brother, now beyond exhaustion, sinks down onto the cave floor. I bring him water, encouraging him to take long sips. And as he falls into a sweaty, restless sleep, I sprinkle his face and forehead with drops of the cool water; I do this for as long as my own strength allows, then I too drift off into a sleep filled with throbbing, unrelenting pain.

I awaken a short time later to the dim flicker of the oil lamp, look over to see that my brother is still curled up next to me, his torn leg, like my own, is now swollen and oozing a yellow-green discharge. I know the time for both of us is growing short. Though dizzy and weak with pain, I push again to my feet and head out to the cave's entrance. Maybe I can see something I hadn't seen before, or maybe I'll look off in the distance and see our band of warriors slowly approaching. But as I make it to the entrance, I am terrified to find a quickly building wind and dark, angry storm clouds advancing upon the horizon. And upon seeing the flashes of distant lightning illuminating the sky, I know that our warriors will not be traveling far on this day.

I turn and head back to my brother, only to find him again upon his feet with brush in hand, his long thick hair matted with sweat, his complexion yellow and waxy. He paints the sun, the split tree, but this time, from one of the unbroken branches of the tree hangs the fox, dangling upside down caught in a rawhide noose. I move to the oil lamp, hold it up for him as he proceeds to paint a man standing next to the trapped animal, a club made of mammoth bone clutched in one hand.

The next scene under the sun is that of the man walking with the dead fox, unaware that behind him is another man, spear drawn and aimed at his back. My brother, panting now for breath, rushes on. The next scene is of the man, again under the sun, the fox now gone from his clutches, lying dead upon the earth, the spear protruding from his back. And surrounding him are animals of all kinds—vultures, hyenas, wolves, and cougars.

As my brother moves to the next open space, I notice that his images have taken up almost one entire wall of the cave, and there is only enough space left for maybe two or three more paintings. I watch as he now paints the worm-covered, skeletal remains of the man, the bigger animals now long gone. Then he moves again, redraws that same sun, then commences upon a painting that shows only a barely distinguishable trace of the man's skeletal remains, and mostly only his skull at that. The worms, as he paints them, are now returning to the earth, strands of grass growing up around them.

Spent and clearly sinking toward death, my brother leans for a long moment against the wall, then begins again. He paints the sun and that same piece of earth with just remnants of the man's skeleton and skull remaining, a single worm pushing up through the soil near where they lay. He then paints grass and bushes and trees, concluding with another small bird flying in sharp descent just above the tree line, angling toward the earth.

His work now done, my brother lets the brush fall to the ground, then sinks down beside me, his entire body sighing. I reach for the bowl of water, which is now nearly gone, and let him drink all but a little of it. Pushing to my feet, I hobble over to the spoiling bison carcass, now surrounded by flying insects, and pull it over to us. There is only enough meat left for each of us to have a few sparse bites; I pull off a strip, hand it to him. He tries to take a bite, but cannot chew or swallow; he only heaves up a yellowish phlegm. I help him as he lays back, murmuring loudly now, finally relinquishing himself to the pain. Helpless, all I can do is sprinkle water onto his searing forehead, watch and wait during the next few hours as he drifts in and out of consciousness.

As I lay next to him, my mind, though unfocused and agitated, circles the realization that I will soon be alone. From my maze of pain, I can feel the fear, the uncertainty, and the aloneness of death seeping into me, followed closely by a sense of raging panic—a panic tempered only by the diminishment of my strength. With one hand upon my brother's forehead, I find myself peering, in vain, out toward the cave entrance. But there is nothing there, and I know, deep down, that there will continue to be nothing there, except maybe for an animal of prey in search of easy quarry.

I settle back, close my eyes and, while stroking my brother's clammy skin, realize that besides my time of birth, and now at my time of death, I have never been without my brother. In all our travels, in all our battles, our tribe, some fifteen families strong, had always remained as one. And my brother, for as long as I can remember, was always one of the central figures. He could paint, he could carve, he could hunt, he could do battle and, most provocative and frightening to some, he could sit, unmoving, for many, many moments and listen to the silence. It was because of this ability, more than any other, that many in the tribe felt he had special powers; they believed he was speaking with the gods when immersed in the silence. I secretly often wondered what it was that he found there, but was always too frightened, like the others, to approach him when he was there. Except, that is, for that one time when he approached me. I remember that I was a boy, just days before my initiation ceremony into manhood, when my brother sat down beside me before an open fire out on one of the rocky lodges of our former homelands. He stared intently at me for a long moment before he closed his eyes to drift off into the silence. I knew that I could not move, could not make a sound, so I too closed my eyes and sat perfectly still. I could hear everything—the sounds of the crackling fire, the screeching sounds of children playing, the echoes of ivory hammers cracking against stone, and the vibrations of footsteps. I continued to listen, hearing it all, but now finding myself moving from one to the other, listening to each exclusively and carefully until another caught my attention, then I would drift with it for a little while. First came the hammer, then a little boy wailing playfully, then the crackling fire, then a person walking, and on and on it went. And then, suddenly, I began to hear something else. In the background of each sound, and in between each sound, and indeed, *within* each sound, I began to find a very slight and subtle hum, almost like a buzzing sound, very distant, very unobtrusive. Then I listened to the silence itself, and it was there too. Was it, I remember wondering, everywhere? And had it always been? I recall listening for a long moment then, suddenly and without warning, my body began to shake with a kind of terror that I had never before known. And when my eyes flashed open, I found my brother intently staring at me, his gaze somehow telling me that he knew where I had been. He gave me a slight, reaffirming nod, then pushed to his feet. And as he turned away, I vowed never again to return to the silence, because in the silence, I was now certain, was where the gods dwelt.

But now, still staring out toward the cave's entrance, I feel the silence closing in on me. I glance over at my brother who is, for the moment, lying still and peaceful. Even with our certain fate, he had never seemed fearful or panicked, only determined to cover that one wall with his work. Now he was allowing himself to freely drift off, his allotment of energy completely exhausted; nothing had gone to waste. Strangely, I find myself feeling a sudden anger and resentment toward him. It all seemed too easy for him, even with the pain. In an attempt to gather myself and to hold back the overwhelming fear, panic, and pain I feel setting in, I pull in a deep breath, but tremble nonetheless. Huge tears well up in my eyes as my leg now feels like a throbbing piece of ember. I am petrified, petrified like a little child lost and all alone in the woods, watching as a pack of wolves, snarling and hungry, encircle him. Like the child, everything I now face is the unknown, and I am terrified beyond all sense. My mind begins to race, and just as it begins to reel out of control, I feel my brother's hand suddenly move into mine, and squeeze gently. I look up to see him staring at me with startling clear eyes, and he nods. And in this moment, I know that the last of his energy has been saved for me. As I watch him, he closes his eyes and gently smiles. Then, with another slow nod, he reaches up and runs soft fingertips over my eyelids until they too are closed. After several moments in this darkness, and feeling his warm hand in mine, my entire body steadies underneath me. And as I had watched him do so many times before, and know that he is doing it at this moment, I let my breath move on its own—easy, steady, and natural. Miraculously, after only a few moments, my mind begins to calm. At this point, I feel his hand gently tighten its grip on mine, but instinctively I do not open my eyes, but instead drift deeper into the silence, almost as if the warmth of his hand was instructing me to do so. I drift, hearing everything—the distant thunder, the whistle of the building wind, the faint crackling of the oil lamp. I listen to it all until, in the flash of a moment, I find that dissonant buzzing sound, that low, unobtrusive hum that pervades it all. I feel my body relax even more as I open myself up further, floating with it, moving all the while closer to it. And as I drift, my mind flashes back to a time long ago when I was just an infant, an infant playing at the feet of my mother right before she had died. The sound of the silence, I am now certain, was there too. I think of my first successful hunt as a boy and know it had been there as well. I think of the time of my mother's death just after the snake bite, and know it was with her then. And when she had been born, it had been there too. And when my father had been born. And when he had died. It had been with him all the while. And with our tribe's traveling warriors, and with our waiting women and children, it is there. I see the sun, the dark clouds, feel the raging wind. Is the sound of the silence there as well? A part of it all? At all times?

My eyes flash open. The churning of my mind has brought me back. Taking a deep breath, I realize that I feel no pain at the moment, and that my mind is steady and clear. I look over at my brother, his hand now removed from mine, and see that he remains still and relatively peaceful. Then, hearing the sound of pounding rain from outside the cave, I push to my feet and hobble back to the tent, searching desperately for the two bowls I remember seeing here. Finally, I find both lying underneath a fox skin. I scoop them up and, as quickly as I can, struggle to make my way to the cave's entrance. I know that in a very short time, I will no longer have the strength to make this journey; this might indeed be my last opportunity for water.

As I move, I feel myself doing no more than dragging my dead leg. If it wasn't for this throbbing ache, I would have no feeling remaining from my thigh down. Regardless, I limp my way through the cave's jagged incline, the tiny rocks and pebbles tearing into my flesh as I pull my foot along the floor. My strength faltering with each and every step, I eventually make it to the outside, slumping down on a large rock that lies just beyond the mouth of the cave. Panting and dizzy, I let the cool rain pour over my body realizing, more than at any other time I can remember, just how good it feels. I vaguely wonder if it had always felt this good, but I just had never before noticed it? I pull in a deep breath, drop my head back to let the water splash down upon my face and eyes. The wind and thunder I had heard just a few minutes before seems to have dissipated for these few minutes, and the rain, thick and heavy, was falling in straight torrents. I look out over the forest below, and then at the gray horizon that seems, all of a sudden, to blend indecipherably into the swollen sky, and marvel at how beautiful it all is. Has it all always been this beautiful? Closing my eyes, I find myself wondering if the sound of the silence is there as well? I breathe in, and as I ponder the question, I notice how the rain pounds at my bare skin, cleansing me. And as I begin to revel in each and every drop, a surge of lightning suddenly breaks the sky, sending a growl of thunder across the horizon. Hearing this, and heeding its warning, I quickly fill the two bowls with the falling rain, pull myself up, and carefully limp back into the safety of the cave.

When I finally return to where my brother is lying, I find him awake, his gray, weakening eyes intently following me as I ease down next to him. Knowing that he is unable to drink, I pour the cool water into my palm and gently rub it over his forehead. When this is done, I then sprinkle a little onto his dry, parched lips, motioning for him to drink. He sighs, closes his eyes, and I watch helplessly as he again begins to fade in and out of consciousness.

Exhausted, and still dripping wet, I lay my body down near his, holding my lifeless leg as still as possible. Feeling myself quickly drifting off to sleep, I know that when I awaken, I will crave food; the bison carcass is picked clean, and no other source of food is available anywhere that is within realistic reach for me. Along with my brother's condition, this is my greatest concern as I slip behind a deep wall of sleep—a sleep I know will be filled with incoherent, violent dreams.

◆   ◆   ◆

Having no idea how long I've been asleep, something, an unfamiliar, alarming sound of some sort, suddenly awakens me. Sitting up, off-balance and dizzy, I listen closely. What had I heard? Had it just been a part of a dream? Afraid, I look around, the dim flicker of the oil lamp throwing dancing shadows upon the painted wall bringing, if only for a brief moment, my brother's images to life. With my heart racing, I take in a deep breath to calm myself; and just as this strange panic begins to subside, I hear, coming from deep in my brother's chest, what I know is the sound that had shaken me from my sleep. I look down at him, only to find that his lips, now a light shade of blue, are oozing a grayish foam. Panic sets in again as I quickly slide over to him, finding myself staring down, first at yellow, waxy skin, and then into gray eyes that are open, but blank and unseeing. At this very moment, I know that he is gone from me.

Desperately, I lean down close to him and listen, but hear no breathing. I lay my hand lightly upon his chest, but feel no movement. I reach for his hand, but it is cold and rigid. With tears suddenly pouring from the corners of my eyes, I push back away from his body, wrap my head in my hands. Then I feel myself rocking back and forth, vaguely wondering if those deafening cries I am now hearing are my own. When I remove my hands from around my head, I realize that indeed they were; and as the echoes fade and the silence again resumes its rightful position, I know with perfect clarity that I am now all alone.

The next few hours move slowly for me. In my despair, my hunger has momentarily faded, but I know it is only a matter of time before I will be desperate. If I had the strength, I would cover my brother's lifeless body, then perform a tribal ritual to honor and prepare his passing, but all I find myself doing is sitting next to him, helpless and staring. He was a part of the gods now—something I always felt he had already come to achieve even in life. I reach out once again to hold his cold, rigid hand, and let my eyes close. His secret, he had made clear to me, was found in the silence.

As his eyes had instructed, I begin to follow my breath, cool as it enters through my nostrils, warm as it exits. Feeling my now infected body relax underneath me, I let myself slip into the silence, finding that subtle hum almost immediately. Still holding onto his hand, I drift with the gentle, encompassing sound, knowing it is everywhere, and knowing it had been in existence long before I ever came to be, and would continue long after I was gone. Maybe my brother had come from it? I hear myself thinking. And maybe he had now returned to it? I continue drifting, letting my thoughts come and go as they wish. I think of the glorious forest below and of this sound of the silence being there right now, and how it was there every day before me and will continue to be there every day after me, then I let it go. I see the horizon, hear the silence, then I let it go. I see my mother and my father, hearing the silence with them, then I let them go. I see the warriors climbing a mountainside, then I let them go. I see the beautiful girl of the tribe who shared her tender young body with me, and who longed for a child, then after hearing the silence, I gently let her go. I think of all the mighty animals roaming the open terrain, hear the silence, then let them go. I see my brother, see his painting with that same sun overlooking it all, and then I let them go. Finally, all thoughts seem to merge into one, and my mind, my awareness becomes as clear as standing water, leaving me with only the sound of the silence. Is this where I am to return? This question is the last conscious thought I have as I drift away into perfect emptiness.

◆   ◆   ◆

... Can I return to what's never left? As before, the sudden churning of my mind has brought me back, this strange, unprovoked question mysteriously upon my lips. Feeling an instant and crushing hunger, I remove my hand from my brother's and quickly push to my feet. But after standing for no more than a moment, I have to drop back down because of a lack of energy and strength. Something, I can now feel, is attacking my body, slowly eating it away. I cannot give up, I tell myself, I must hold out for as long as I can. But to do so, I must eat. In what I know is in vain, my eyes scan the cave, only to find that nothing has changed. Feeling my energy waning even further, I lie down on my side, curl my upper torso downward so I'm lying in a half fetal position, and rock myself to sleep, but find that I am only able to sleep in spurts. The hunger jolts me awake each and every time. And each time I awake I sip at the water, but it doesn't help. Finally, after many hours pass, my head grows light and dizzy and I feel myself descend into something that resembles sleep, but is not quite sleep. All the while, a troubling idea circles through my mind as I fade in and out of this delirium.

◆   ◆   ◆

I awaken some time later, my hunger cruelly permeating every part of my body. With all my strength, I sit up, stare over at my brother's corpse, now slowly creeping toward the first stages of decomposition. If I do not eat, I tell myself, I may not make it through the night. Without another thought, and knowing I have both no choice and virtually no time remaining before my strength vanishes altogether, I reach for the large knife sitting next to the bowls of pigment. With all I have left, I push myself up so that I am straddling my brother's body. My mind is dizzy, cloudy and fading; I could almost believe I was dreaming. Working at a fevered pace, I feel for the area of his chest just above his heart, raise the knife up over my head with both hands and send it down in one great thrust that pierces the very center of his upper torso. Then, with the knife implanted, his blood having squirted onto my face and arms, I begin to carve out his heart, tears of despair and confusion draining down my face.

In a matter of five minutes, I have the large organ laid out before me, his bright red blood draining out into puddles all around me. Without hesitating any further, I begin to greedily consume my brother's thick, strong heart, my battered body moaning in relief with each bite that is ingested.

Once I am filled, I take a long drink from the closest bowl of water, and then lie back from exhaustion, leaving his blood drying upon my bare skin. I know I will lie in this same position until I fall into a deep, agitated sleep. But before I do, the thought keeps echoing through my mind that something is now actively killing me. I haven't looked at my leg in more than a day, but the last time I did, it was swollen and bright red, and the wound itself was green and full of puss. Pushing it from my mind, I close my eyes and drift off in a search of the silence.

◆   ◆   ◆

I've just awakened, but I can barely move. My head is burning with fever, my lips are dry and cracked, and my entire body is now aching beyond all tolerance. How long I have been lying here, I do not know; it has been at least a good part of two full days and nights. All sense of time has left me. My eyes are open for the first time in hours, and I notice that the oil lamp has burned out, leaving me in complete darkness. I close my eyes and return to the silence. I am not frightened.

◆   ◆   ◆

I've awakened again. I hear what I think is thunder rumbling in the distance, and heavy rain. My eyes, now seeping, remain closed, and I drift away again while seeing flashes of my brother's last painting as I go, vaguely wondering if the worms have found his body.

◆   ◆   ◆

I may be awake, but I am not sure, for all is muddled. My breath is heavy, and coming more slowly now. I drift, fighting for air, my lungs weak and fluttering. I hear the silence with perfect clarity as I pull in one last breath that falls just short. But I feel warmth, and all is bright, and so very gentle. And I hear soft whispering, beckoning me …

(21ˢᵗ century)

Rudy—

Along with a bunch of other things, I found the story you've just read in his files. He called it *Mistress of Time (Chapter 1)*.

James

Cc: Kevin and Nancy

1

March (Morning)

They could hear the sound of the river fading away behind them as they pushed up the wooded trail toward the rocky overlook. The morning air, though unseasonably warm for the end of March, still bit at exposed skin. Ears, nose, and lips glowed a bright red.

With wooden staffs digging into the dirt with every other step, the two brothers walked briskly, one behind the other, up the steady incline. The section of the woods they now traveled contained mostly oak, ash and hickory, and would have provided, if this hike had been two months later, a full canopy of shade. Today, however, a gray sky was visible from all angles. Rudy, who was several paces ahead, called back to his brother. "Let's go to the left," he motioned ahead to an upcoming fork in the trail. "There's a nice rocky overlook up there. It's close to the parking area where Sinclair's going to be dropped off to meet up with us."

"Sounds good," James agreed.

The new direction of the trail provided a bit of a break in the tree cover the further up the hill they moved, though the angle of incline grew by substantial measure. They pushed forward in silence, the flow of the river receding with each step until it was just a distant hum in the background. When they finally reached the opening in the woods where the hill began to level as it yielded to a series of large rock formations, they made their way to one of the flatter formations. This location provided not only a nice place to sit but also magnificent views of both the Patuxent River below and a large expanse of the Patapsco Valley State Park.

When they finally reached their chosen rock platform, both brothers bent at the waist, their gloved palms resting on their thighs, and they breathed heavily from exhaustion. "Quite a climb," Rudy said after a long exhale. He stood, wiped the sweat off his forehead with the back of his glove, smiled at his younger brother of three years. "You doin' okay, young boy?"

James, still bent over, looked up out of the corner of his eye, gave him a crooked smile. "I told you I was out of shape."

"You weren't lyin'," Rudy laughed as he leaned his staff against the rock and pulled his backpack off. Unzipping a side pocket, he pulled out two bottles of water, handed one to his now standing brother.

"Thanks," James said with a final exhale, looking out over the view that spread out before them. "Wow," he said as he laid his staff across the rock and took a long gulp of water. His gaze swept across the horizon until it came upon an opening in the tree line that exposed the river below. "Very nice," he nodded with approval. "Very, very nice."

"I say we rest here for a while," Rudy suggested.

"Sounds good to me," James agreed as he swung off his backpack, and carefully eased his tired body down onto a flat area of the rock. Rudy sat down next to him, took a long, deep breath and then held up his bottled water in a silent toast. James extended his water in return. "To his memory," he said quietly.

Rudy took a long drink while, almost subconsciously, reaching under his sweatshirt with his free hand to pull out the twenty inch silver chain and the attached silver cylindrical-shaped pendant that was around his neck. For a long moment he let his fingertips gently trace the etchings on the pendant before letting it go to dangle freely from his neck. He took another deep breath, exhaled slowly, and stared out at a gray sky that was now slowly yielding to an almost turquoise blue. "Yes," he nodded finally, "to his memory ..."

The brothers sat in silence, taking it all in. This was the first hike they had taken since the death of their father two months prior. They had vowed to one another the day after their father's memorial service to pause from their busy lives one Saturday every month for a long hike. They also vowed to, as much as possible, discuss during those Saturdays only those things that their father cherished the most. The things he said were often left behind during life's mad scramble, the things that gave life its "texture, its feel, its underlying meaning".

Rudy let out a long sigh, glanced at his watch. His twelve-year-old son, Sinclair, was due to be dropped off by his uncle in exactly one hour. Sinclair had spent the night at his cousin's house, who happened to live near the park.

"You suppose Dad was right?" James asked suddenly.

"About what?" Rudy looked at his thirty-five-year-old brother, who was holding his own identical silver cylinder-shaped pendant between his fingertips, staring at the etching that resembled the number eight.

"About infinity," James answered, dropping his pendant so it swung freely on his silver chain.

Rudy looked out over the horizon. "All I know is that he talked a lot about it. He truly believed that the universe is infinite. He talked constantly about what the observations made by Hubble seemed to show—that the universe was expanding … and that the expansion seemed to be accelerating."

"And if space is infinite," James nodded, recalling his father's words, "then so too is time." He smiled. "I remember him saying that since Einstein had added time as the fourth dimension, thus linking space and time, then if one is infinite, so too must be the other."

Rudy laughed. "That reminds me of just how much he loved Spinoza's *sub specie aeternitatis*."

"*Under the aspect of eternity*," James translated. "If you think about it, it makes a lot of sense … the notion of viewing yourself and your life through eternity's gaze—especially if you consider the average life span, at least in the industrialized world, is roughly eighty or so years. If eternity means forever and if forever actually does exist, as an infinite universe would suggest, then place our eighty years against that backdrop."

Rudy smiled. "Not long, is it?"

"To time," James shook his head in return, "it's not even a wink."

"Certainly gives you a different perspective," Rudy agreed.

James pulled a scarf out of his backpack, wrapped it around his neck. "It makes me realize that we must be a part, though small it may be, of a progression of some sort. But this seemingly miniscule role we play in the overall scheme doesn't diminish it for me. In fact, it enhances it."

"In what way?" Rudy wondered.

"Well, it allows me, in a sense, to look from the outside in. And I see the grand, overall structure of time and all that it encompasses, and knowing we are a part of such a grand structure that may truly be without end brings a sense of enhancement. Without this perspective, I find myself getting trapped within myself, looking from the inside out, and I lose the sense of the grand scale in which we exist. It is during this self-containment, this closed-in perspective that I feel tied to the fleeting whims and passions of being human. Seeing myself through the eyes of eternity brings a sense of connection that is not there otherwise."

"Some would say just the opposite," Rudy pointed out. "They would say that this perspective makes them feel painfully insignificant—makes their lives virtually meaningless."

James shook his head. "I know … but that may go back to a kind of conditioning we've experienced as a species for so many centuries. At some point in time, we came to believe that we should hold a kind of preeminent role in the overall scheme, and we began to define this role in human terms. The wraths and blessings of Nature, we came to believe, were in direct response to our behavior. Our perceptions of the gods mirrored our own tendencies, our own likeness, and the story of the universe, from its beginning to its end, rested upon our actions or inactions. How long have we been told that the world will end when we've completely lost our way, when we finally stray fully from the intended 'path'?"

"The End of Days ... and such things," Rudy added.

"Exactly," James nodded. "And that way of thinking, of approaching the world became engrained in the human psyche. So engrained, in fact, that we can no longer comprehend how beautiful and spectacular it is to be a part of a greater whole. We feel diminished by it when we should feel more complete."

Rudy shrugged, thought for a long moment. "Our belief systems, in a sense, became cathartic. They became a way for us to cope instead of a way for us to truly understand the world in which we exist."

James nodded. "And the need for comfort soon outgrew our need for truth, and when the possibility of truth does present itself, we struggle to not only comprehend it, but simply to recognize it, and to cope with this recognition."

"Almost like we're imprisoned by our beliefs," Rudy added, "by our systems of thought."

"It indeed seems so," James agreed as he laid back onto the rock, his eyes staring up into the clearing sky. "What a great day it's turning into."

Rudy nodded in agreement. "For March, it's incredible."

The brothers sat in silence for the next several minutes. Rudy closed his eyes, and let the distant sound of the river find him. The things his father had said over the course of their years together seemed to have flooded back to him in very vivid ways since his passing. Taking a deep breath, he concentrated on the flow of the river, and heard his father's voice quoting once more the ancient Greek philosopher, Heraclitus, "Son, always remember, you cannot step into the same river twice." Rudy remembered being about seven years old the first time his father had said this to him. At first, Rudy looked at him askance, thinking in seven-year-old language—"why of course I can, Dad". As if reading his mind, his father smiled down at him. "No you can't, Rudy. Think about it."

And Rudy did think about it, but it never came up again until a year later when Rudy was feeling a crushing sense of loss because one of his best friends was moving away. Driving home from his final goodbye to his friend, his father reminded him about the river. "You recall what I said about the river, Rudy?" Rudy nodded, the tears that had long welled in his eyes, now poured over. And he understood.

Rudy opened his eyes, looked over at his brother. "What do you think about when you hear the river?"

James laughed, not needing to answer.

Rudy shook his head with a smile. "Well ...?"

James sat up. "It seems pretty evident now."

"Sure does ..." Rudy agreed, "everything's in a constant state of change ... his death, for us at least, serves as ultimate verification of that."

"Nothing's static, that's for sure," James added, "Nature seems to tell us that pretty clearly."

"But do humans tend to have, at least in our belief systems, a tendency to gravitate toward the static?" Rudy wondered.

"Whatya mean?'

"Well," Rudy adjusted himself, pulling one leg up toward his chest as his mind worked over what he was about to say. "Such as the notion," he began slowly, "that all things are in a constant state of change in this existence, but that they are somehow static in another realm."

"You mean the notion of Heaven and Hell, eternal life, etc ...?" James asked.

"Precisely ..."

James shrugged, stretched his legs out before him. "What you're getting at here is the age-old philosophical discussion regarding if there is, in the end, an ideal realm, a universal, unchanging structure to everything. You know, back to the days of Plato and the boys."

"Well," Rudy continued in the same careful tone, "the notion of eternal life—Heaven, Hell, and the like—implies that there is some universal, unchanging realm behind it all—and not just for that which is physical, but also spiritual, moral, etc ... There is, if such a notion is correct, an ideal version of everything that exists, and everything in this earthly realm is a mere representation, imperfect at that, of the ideal form ..."

"*Form* is actually the very word Plato used to describe the notion," James added, "not so much Heaven and Hell, but that of an ideal perfect realm."

"And if this all is correct, then there's everything from the perfect, ideal flower to an unchanging ideal moral code present in some transcendent realm, and we are striving, at all times, to reach this ideal realm, but we can never quite get there in this existence."

"The best we can do is to reach mere representations of the perfect flower, the perfect moral life, etc ..."

"The question is: is there a perfect standard of, let's say, justice? Morality?" Rudy stared at James directly.

"If there is, then there can truly be no consideration given to relative positioning. No gray areas."

"Let's take morality," Rudy suggested. "For example, lying. Lying is wrong. But are there any cases where lying can be justified? How about," Rudy thought for a moment, "someone with a gun is looking for your best friend to shoot him, and asks you if you know his address. If lying is wrong, then you would be obliged to tell him the address."

"Or," James held up a finger, "you can say: 'I know his address, but I'm not going to tell you because I know you intend to do him harm'."

"Indeed you could," Rudy agreed, "but let's take it to the next step. The gunman tells you that he will shoot you in five seconds if you don't tell him the address. For some reason, you have a feeling that if you give this deranged gunman any address, even a fake address, he will leave you alive and well. At this point, you have a choice—die or give him an address. Would a person really be expected to give up their life in such a scenario over the moral principle that it is wrong to lie? Or would it be reasonable for someone to conjure up an address, get rid of the gunman, and then go call the police?"

"It seems pretty obvious what the answer is."

"But if there is an unchanging, ideal moral code, then it is binary in nature."

"Meaning what?" James asked.

"Meaning, if there's an ideal moral code dealing with lying, then to lie is either breaking it or not. There's no gray area allowed in such a structure."

"So the person who gave the killer a phony address would be guilty of breaking the transcendent, ideal moral code that 'lying is wrong'."

"And if there's a God making judgments based on one ideal moral code, as concepts such as Heaven and Hell suggest, then that person has fallen. Even if having given the killer his friend's actual address would have very likely meant that his friend would have been murdered."

"But Rudy, many people would argue that God would understand the situation and judge accordingly."

Rudy nodded. "I understand, but that seems to be having it both ways. How can you say on one hand that there is one realm that is perfect, such as Heaven, and one that is imperfect, such as the earthly realm, and then allow a gray area to emerge based on circumstance?"

"One might argue that God would understand the dilemma, recognizing that precisely because it is in the earthly, imperfect realm that such a thing could happen. In the perfect, heavenly realm such a situation would not have occurred in the first place."

"Because all is perfect?" Rudy asked.

"Correct."

"But we are to be judged based on how we respond in this life?"

"According to the faithful, by free will," James reminded him. "And tests …"

"But the question is still not answered: would it be ultimately wrong, in God's eyes, to lie in that situation? If not, then there's a huge contradiction lurking. What this is saying is that what's correct in the earthly realm is not correct in the heavenly realm, we understand this. But if *God himself* grants you entrance into the perfect realm of Heaven, in essence saying it is *ultimately* all right that you lied under those circumstances, earthly or not, then isn't he conceding that there is in fact a distinction to be made? If this is so, then there's not one ultimate standard that dominates the universe, because if the universe is God, as many faiths claim, then what is being said here is that God, and his universal order, is somehow divided, relative, and nuanced."

"Either that," James offered, "or God is saying: 'yes, what you did is wrong, but you're forgiven', and thus will be allowed to enter into the perfect, heavenly realm. The faithful would say you have been forgiven for your act, not that your act was okay."

"But I should be forgiven and allowed into Heaven based only on if I repented for my wrong actions, but I wouldn't feel any remorse, nor would I be about to repent, because I wouldn't think I did anything wrong. With this said, the question remains, would God himself consider my lie in the given scenario wrong?"

"Maybe it would be understood in the context that you committed the immoral act in the imperfect realm, which put you in that position precisely because it is imperfect'."

"But if I'm still allowed entrance into Heaven, and I haven't repented, then God is saying that my act was ultimately okay because of the context in which it occurred. Wouldn't this suggest that one, underlying standard of right and wrong does not then exist?" Rudy paused for a moment, gathered his thoughts before continuing. "And on top of it, I'm assuming that murder is also one of those universal morals. Murder is wrong. If this is so, and I don't lie to the gunman, then I'm knowingly allowing a murder to take place. Is that not wrong in and of itself?"

"The answer would probably be no," James offered, "because you didn't actually commit the act of murder yourself."

"Under that logic then," Rudy pointed out, "Judas wasn't guilty of an immoral act, because he didn't commit the actual murder of Jesus?"

"Well, that's entirely different."

"How? If I would have given my friend's address to the killer, wouldn't I have betrayed my friend in some way? And isn't betrayal itself immoral on some level?"

"But you weren't betraying him out of greed and avarice, like Judas betrayed Jesus."

Rudy looked at him crookedly. "So now there's a relative position in regard to betrayal. Under some circumstances it's understandable, under others it's not. It comes down to the motivation behind the act, and the context upon which it occurs. Doesn't that, in and of itself, serve as just one more instance that undermines the 'there's one underlying moral standard' argument? Lying is not *always* wrong, and neither is betrayal *always* wrong."

"Well there's definitely a difference," James pointed out, "between greed and self-preservation."

"But there's a distinction being made nonetheless, and therefore an undermining of the entire argument that there's one moral standard of right and wrong, truth and justice, etc … After all, James, we could come up with a million scenarios, from murder, to lying, to stealing, to cheating … with a thousand different distinctions, relative positions, etc … within each and every one."

"And thus a constant undercutting of the notion that there's a perfect, unchanging ideal to all things," James conceded.

"And it's not just in religion," Rudy quickly pointed out. "But also, to a lesser degree at least, in science as well."

◆   ◆   ◆

"Science?" James asked curiously.

Rudy nodded. "If you look at the Scientific Revolution of the 16th and 17th centuries, for example, and the subsequent dominance of Newtonian physics, did we not, through the resulting notion of a clockwork universe, attempt to create a static picture of our world? After all, it was thought that through our knowledge of the laws of motion, we could accurately chart, through cause and effect, what would likely happen in the future given any initial action. Again, was this deterministic, mechanical view of our universe not in itself a natural desire to gravitate toward the unchanging, the predictable?"

James shrugged in thought. "In a sense, what you're saying is that religion and science seem to be naturally, though maybe subconsciously so, searching for the same end point?"

Rudy nodded. "A greater understanding of the world in which we find ourselves, and an understanding we are capable of grasping—which inevitably, it seems, always tends to want to move toward a static picture."

"Because this is the condition that seems to give us the greatest level of comfort," James followed, "the thought of a world that allows change within itself, but yet does not itself change. We like the notion that the stage is always there for us—a stage that is always in the same location, possessing the same dimensions, the same lighting and acoustic capabilities …"

"But a stage," Rudy added, "in which we are free to move around the props, change our costumes, adjust the lighting and acoustics as need be, and play out our different roles. The thought of a stage in which change is part of its own fundamental condition frightens us … frightens us because suddenly the grand stage does not seem to be there solely for our ends."

"And thus we are just one of its many backdrops, not the focal point of its very existence," James concluded, then paused, and studied his brother. "But you said science, though it too gravitates toward the static, does so to a 'lesser degree' than religion. What did you mean?"

Rudy pulled in a deep breath, thought for a long moment before he answered. "The subconscious human impulse, even in science, naturally tries to seek out a condition that is within our realm of understanding—and as we've been saying, a reality that is static in nature tends to fall within the human comfort zone. But science, unlike religion, has certain safeguards in place, such as experimentation and verification. In the world of science, if a theory cannot stand up to the stringent tests of experimentation and quantitative verification, science will move on. Religion, on the other hand, has safeguards in place, such as faith, ceremony, ritual, and the like, that work in just the opposite direction. Religion's safeguards tend to stymie any potential movement away from the static, away from the comfort zone, regardless of any evidence to the contrary."

James nodded thoughtfully, unzipped the side pocket to his backpack, pulled out two small bags of peanuts. "But the people of science also battle the tendency to move toward the static?"

Rudy nodded. "It has to be a constant battle …"

James tossed him one of the bags. "Hasn't the quantum revolution really challenged this tendency … in a sense turning the whole apple cart upside down?"

"It indeed has," Rudy nodded as he peeled open the bag. "Actually, this deterministic vision put forth by the Newtonian revolution began to show cracks as early as the mid-1700's. A good starting point to show the evolution away from determinism is James Clerk Maxwell's work with electromagnetism in the

1860's. As you know, electromagnetism essentially became another branch of physics to accompany Newton's mechanics."

"But shouldn't we start first with Michael Faraday?" James asked.

"We could actually," Rudy agreed, "especially given Newton's view that universal gravitation was caused by a kind of action-at-a-distance—basically asserting that some invisible force reaches out directly from one object to another through empty space to cause the subsequent pull to occur. And you're right, Faraday, through the principle of induction in 1831, brought us a better understanding of what became known as field theory, which better explained various phenomena, such as Newton's gravity."

"Absolutely," James continued the thought. "In Faraday's time, one of the big questions of the day was the behavior of electricity. Faraday himself was insistent upon the notion that electricity did not flow through a wire in the same way that a fluid might flow through a hose, but instead sent out lines of force *beyond* the wire. This was the reason he proposed why a magnet that was set outside an electrical wire moved. This view essentially placed him at odds with the established scientific community of the day."

Rudy nodded. "And when this was first discovered—that a magnet moved when placed next to an electrified wire—scientists were perplexed."

"Largely because they held on to the notion that electricity flowed in a straight path down a wire," James pointed out. "The thought that lines of force were created as electricity flowed never occurred to them. What's more, Faraday also believed that these lines of force could actually flow *around* a wire. And before long, he further surmised that a magnet emitted the same kind of lines of force, and thus these two forces, the magnet's force and the electricity in the wire, interacted with one another, thus explaining how one seemingly affected the other."

Rudy smiled. "I had almost forgotten all of this."

"Me too," James admitted as he sat forward. He narrowed his eyes as if to stretch his memory a bit further. "Remember how Maxwell eventually took this all a step further by illustrating that a static magnet could move a wire. This was a huge discovery in the realm of science. Just practically speaking alone—this led to the creation of the first electric motor."

"If I remember correctly," Rudy contributed his own recollection, "it would also eventually prove to be significant in helping to better explain, among other things, gravity."

"Correct," James nodded, "no longer did one object stretch out across empty space to exert a pull on another object, but instead each object created its own field that interacted with the field of any other objects within its vicinity. And further, an object's field would not instantly disappear if somehow the object itself disappeared … basically asserting that the field has, in essence, a physical existence of its own."

Rudy nodded, dropped several peanuts into his mouth, and looked out over the horizon. "Faraday was indeed important … not only had he created motion between electricity and magnetism, and all the practical implications it brought, but he had also inadvertently stumbled upon something that would come out later—the notion that there had to be something common to all motion, and common to all fields."

James smiled knowingly. "Energy."

Rudy nodded. "When you mentioned the physical existence that fields possess—well, energy is the substance from which all the various fields, from gravitational to magnetic to electrical are made."

James leaned back onto his elbows. "Faraday was also interesting in the fact that many of his scientific notions evolved out of a religious framework. He looked upon his scientific works as a way to better understand God."

"Wasn't he a Sandemanian?"

"He was—and his core belief was something he referred to as the 'unity of the forces of Nature'. Since everything was created by God, he believed that fully understanding one part of Nature could possibly lead to a complete unveiling of how everything else was connected to it."

Rudy looked out over the bare tree line below. "Interesting indeed," he said more to himself than to James. "Reminiscent of Einstein and Spinoza." He took another long drink, twisted the cap tightly back onto the plastic bottle. "And you're correct, James, Faraday indeed set the stage nicely for Maxwell."

James leaned back until his head rested against the cool rock. He thought about Maxwell's intense focus on the relationship between electricity and magnetism. Since it was already known that a moving magnetic field created an electric field, Maxwell concentrated on the possibility that a moving electric field, in turn, should create a magnetic field. James was amazed at how much of a natural, and somewhat common-sense proposition Maxwell was putting forward. After all, the inherent balance of Nature would seem to dictate that such a phenomenon should not only happen in one direction.

He sat up. "I'm thinking about Maxwell's equations. How they not only proved his assertion that a moving electric field produced a magnetic field, but how they ultimately served to unite electricity and magnetism."

"Electromagnetism," Rudy looked over at him knowingly.

"And as astounding as these equations were," James continued, "the revelations they led to were equally astounding."

"That a moving magnetic field creates a moving electric field, which in turn creates another magnetic field, which then creates another electric field …?" Rudy asked.

"Exactly," James lifted his hands in amazement, then moved one hand quickly over the other to demonstrate, "and on and on it goes, electric field creating magnetic field creating electric field … shimmying through space one after the other…. and what's more, moving through space as its own entity, carrying energy …"

Rudy nodded with a smile. "And moving through space," he held up a finger for emphasis, "*as a wave* … and this wave, Maxwell predicted, would move through space at precisely the speed of light." He tossed the bag of peanuts back over to his brother. "And thus electromagnetic waves were born."

James caught the bag with one hand. "And so too was the knowledge that light itself was a form of an electromagnetic wave."

2

Rudy—

Here's something else I found. He called it "Quantum Discernment, Observation, and the Merging of Physics and Metaphysics."

James

Cc: Kevin and Nancy

Plato (427–347 B.C.) begins his "Allegory of the Cave" with the scenario that places an individual, or a group of individuals, in a darkened cave. Imagine further, Plato says, that not only have these individuals lived the whole of their lives in this cave, but they have lived it while being chained in such a way that they have been facing only the back wall of the cave. What's more, from the time they were mere infants, a stage had been erected behind them where manufactured objects were continually held up that resembled men, animals, plants, etc ... And behind this stage, from the very beginning, has been a continually burning fire that flashes the shadows of these fake objects on the back wall of the cave. Plato argues that these men, having spent all of their lives facing the back wall and thus seeing only the shadows, would naturally assume that these shadows represent reality. Because of their circumstances and positioning, they would be unable to comprehend that these shadows were in fact what Plato calls *images*, mere ethereal projections, of a more substantial, fundamental reality. Plato terms this state of cognition *imagining* because it is the furthest removed from reality; for this very reason, these objects, according to Plato, are the lowest objects of contemplation. It is in this realm, he suggests, that most people live their lives.

From this point, Plato tells us to imagine that one of these individuals is suddenly released from his chains and made to turn around and face the stage. The first response of this individual would undoubtedly be shock, pain, and fear caused by the blinding light of the fire, a light to which he would be completely unaccustomed. His first impulse would be to turn back toward the familiarity of the wall, and probably would do so if not forced to face in the opposite direction. But being forced to remain, he would soon find that once his eyes adjusted, he would be able to focus in on the manufactured objects that had caused the shadows which he had, for his entire life, thought of as reality. At this point, the individual would enter a second stage of cognition which Plato calls *belief.* Though belief, to Plato, is a higher level than imagining, it still does not represent a "true" knowledge of reality. These objects of belief are still just images of the authentic objects themselves.

Imagine, Plato continues, that this individual is now forced out of the cave altogether. When he reaches the outside world, he would again be blinded, but now by the fierce light of the sun. Again, he would fight to run back into the safety of the cave, but this time, however, when he became calm and his eyes grew accustomed to the light, he would find himself elevated to the next level of cognition, which Plato calls *knowledge.* This level would have been reached because he would now be able to see things as they truly are, not just their likeness. In essence, the light of the sun would allow him to comprehend the true nature of reality. And to take it to yet another level—when he again overcame his fear and his eyes became completely adjusted, he would be able to look into the sun directly, thus enabling him to discover the full realization that the sun is the true cause, the true source of all things in this upper world. At this point, the full and complete realization dawns that the objects and shadows from his life in the cave represented only mere appearances, imperfect representations, of a more ultimate reality.

The "Allegory of the Cave" is, among other things, a powerful representation of the journey of the human spirit from that of *becoming* to that of pure *being*, the latter of which offers forth an individual with the capacity to understand the intelligible world of *forms*.[1] When contemplating Plato's cave analogy, one may find it interesting to examine the motivations of the individual who breaks free from their chains and ventures from the shadows. There are two possibilities through which one can examine these motivations. First, the individual, largely against their will (at least initially), was freed and guided from the cave by some external force. Second, that he or she somehow freed themselves of their own volition and faced the unknown willingly, independent of external influences. From these two possibilities, the duality is born that represents the two foundations from which sentient beings launch their quest toward truth and understanding.

---

1.   To Plato, *forms* are a transcendent reality. All that exists in our physical, sensory world, including the realm of ethics, has a more perfect representation of itself in the transcendent world of forms. For example, "real" waves are imperfect representations of their ideal transcendent form.

The first individual, having found the more ultimate reality (and we assume a greater sense of truth) upon reaching the outer world would, in most instances, naturally pay some level of homage toward whatever external force it was that guided them from the darkness into the light. One would think, particularly if the external force continued to assert itself upon the individual in this new world of light, that the individual would develop a sense of dependence upon this force, not to mention some degree of loyalty.

The second individual, who left the cave of their own volition, would develop a wholly different sense of not only the reality discovered, but of the journey itself. To whom would this individual pay homage? Where would his or her dependency/loyalty lie? We can envision an individual becoming dependant solely upon his or her own instinct and reason, in conjunction with a continually growing, trial and error-based knowledge of the surrounding world. Though this individual's motivational launching point and subsequent path would be wholly different from that of the first individual, the same "truth", in this case the light of the sun, awaits both. It is at this point that human perspective and interpretation enters the equation. Again, we return to the initial launching point—that of the base motivation of each upon the initiation of their journey. Interpretation and perspective will be dictated by the different journey each individual has experienced to obtain their truth; thus we witness the historical duality which is inherent in the human quest for truth. If the discussion of the motivation of the individual leaving the cave is introduced as a part of Plato's analogy, one could interpret it as being a representation of the duality in the quest for truth—a duality classically represented by the differing approaches put forth by religion and science. The external force responsible for the first individual's quest for truth could be interpreted by the individual as being that of "the hand of God", and thus a journey orchestrated through a religious context. The second individual, through dependence upon reason and the examination of his environment, found a journey that would be more reminiscent of a scientific journey of discovery.

One of the most paramount challenges for the human species is how can this apparent polarization between the two journeys, that of science and religion, of physics and metaphysics, be reconciled? Is there a bridge that can truly be built between the two? To address this question, one can again return to the cave analogy. The focus of the analogy is primarily upon how the individual is affected, first by the journey itself, then by the outside world. But maybe the key is to examine just the opposite, which is: just what effect does the presence of the individual, this new observer who has ascended from the cave, have *upon the outside world?*

The presentation of this question often *feels* counterintuitive to the human mind. We're comfortable in our knowledge that humankind thinks, possesses intelligence, and is therefore conscious. But Nature? Does Nature really *think*? Does it truly have responsive and *discerning* intelligence? Is it conscious in the same way we are? These are questions that sometimes do not come very naturally to the human mind, at least within the context of his or her daily existence. Yes, Nature seems to be intelligent, most will say—after all, it maintains a perfect balance, nourishes life, responds to its own internal conditions. But does it think—in the way man generally interprets *thinking*? Does it have *discerning* intelligence, like humankind, or just a responsive intelligence? And what if it does think, and does have discerning intelligence? The next question would naturally be: where did Nature obtain this ability? Again, we see the duality begin to reshape itself. The world of science will quickly cast a critical eye at the notion that Nature has discerning intelligence, because to concede this would be to take one step closer to the idea that an intelligent force (God?) may in fact exist behind it all. This is an area where most in the scientific community find themselves reluctant to venture.

On the religious side, particularly in the western faiths, the problem with this notion of a discerning natural world implies that man may not, after all, exclusively hold the superior, right-hand position in God's pecking order. The notion of transcendent theology is challenged, thus indicating that man may not, in the end, have dominion over Nature.

What about the idea that the observer, in our case the individual ascending from the cave, causes a reactive response within the natural world? Can the act of observation truly have such an impact? If so, how exactly does this take place? And what exactly is observation? Does it physically exist? From where does it come? And why does it have such a pivotal role in reality? Furthermore, could it prove to be, when all is said and done, one of the portals of entry, the common themes, the basic building blocks from which the uniting of science and religion, physics and metaphysics, can be forged?

It is interesting that Plato's "Allegory of the Cave" bases the ultimate reality upon the discovery of light; in this case, the light of the sun. When studying the origins of Christianity, one understands the paramount role of light in God's creative process. In the world of science, the role of heat and light (energy) in the formation of the universe is well documented. Even in the nondenominational metaphysical realm—with such phenomena as Near Death Experiences, for example—light is commonly reported as being the primary component of the experience. Therefore, it is with an exploration of light that we will, using the realm of quantum physics, particularly in relation to the intelligence of Nature and the role of the observer in the shaping of physical reality, begin our quest to merge the seemingly separate roads of physics and metaphysics.

Does Nature have a discerning intelligence? Clearly the workings of Nature, in the perfect harmony and balance displayed by all segments of the natural world, indicate an intelligence, but is it *discerning*? Does it recognize and react? Is it capable of searching out and choosing other options in a given scenario when the first option is either unavailable or unworkable? Humankind is a discerning creature—humans have the ability to recognize and adjust within a given situation, and thus are able to find alternative solutions when needed. As expected, some animals also have this ability. But what about the realm of Nature—the fundamental realm—that humans generally regard as not being "consciously intelligent"? Does this realm, the more fundamental levels of Nature, possess discerning intelligence? Does light, for example, respond to changing situations in a way that indicates not only consciousness, but an intelligence that recognizes, adjusts, and reacts accordingly? Does it search out alternative options in given scenarios when circumstances require it to do so?

March (Mid-Morning)

Rudy pushed to his feet, stretched his arms toward the clearing sky for a moment, and then looked down at James. "You say we press on, little brother? The parking area is just down the other side of this hill." Rudy glanced at his watch. "Sinclair will be showing up in just about 20 minutes."

James placed the peanuts and water back into his backpack, stood up next to Rudy. "Sounds good." The two swung their backpacks over their shoulders, grabbed their staffs, and began to walk along the rocky trail that led down the other side of the ridge away from the river. A few minutes later, they again found themselves in thick woods, the bright sunlight streaming through openings in the bare branches.

"It's amazing how much light tells us," James said in a near whisper, and wondered, and not for the first time, why one has a tendency to whisper in the woods. "Dad alluded to this in the last essay excerpt I sent you."

"He was fascinated by wave-particle duality," Rudy said as he ducked around tree limbs hanging across the trail.

James paused, unzipped his coat pocket, pulled out several folded pieces of paper and handed them to Rudy. "I found another excerpt from Dad's writings."

Rudy paused, leaned his back against the thick trunk of one of the large maples and began to read:

> ... An examination of the double-slit experiment, originally performed by Thomas Young in the 1800's, serves as an ideal launching point to examine the wave-particle duality of light.[2] The double-slit experiment is structured to observe the behavior of particles (in this case light) in various scenarios. The experiment consists of a light source, an opaque barrier with two vertical, parallel slits, and a photographic screen (or plate). By opening and closing the slits in various combinations, the behavior of the light as it is directed in varying degrees toward the barrier is recorded and revealed by the photographic plate. Essentially what is found is that the photons, which are the quantum carriers of light, exhibit particle properties when placed in certain situations, but exhibit wave-like properties when placed in other situations. This eventually came to be known as the wave-particle duality of light. During the first phase of the experiment, a beam of light, made up of its many bundles of particles called photons, is directed at the barrier with just one slit open.[3] As one might expect, the light goes through the open slit and strikes the screen in one place behind this slit, in essence behaving like a bundle of particles. The light

2.   All particles of matter exhibit both wave and particle properties.
3.   Photons are the quantum carriers of light.

simply hits the screen in a single beam. And conversely, an identical beam is formed when that first slit is closed and the other slit is open. So far so good. Now, what about when both slits are open at the same time and a beam of light is directed toward them? In this scenario, one would assume that some of the light would go through one slit, some through the other, and some through neither as it would be blocked by the barrier altogether. The light that made it through, one would figure, would simply strike the screen behind the corresponding slits, creating two separate beams of light. What happens, however, is that an alternating band of light and dark patterns, called interference pattern, are formed on the screen. These strange, almost ripple-like patterns seem to indicate that the photons are now behaving as a wave. One can think of a wave of water as a more recognizable illustration. What would happen if a wave was rushing at a barrier with the same configuration—two openings? Some water would be stopped by the barrier, but some would get through each opening. One can envision the water that made it through the openings seemingly curling back to, in a sense, interfere with itself. Two waves would have been created that would overlap one another to, in essence, cancel each other out, which would result in a kind of ripple effect. The double-slit experiment indicates that a similar occurrence takes place with particles. In the case of light, the photons seemingly interfere with one another to create the bands of light and dark on the photographic plate.

The next phase of the experiment is set up to ask the question: what would happen if the light is dimmed to the point where only one photon at a time is sent toward the barrier? How would this one photon behave when faced with the same slit options? When one slit is closed, the photon simply travels through the open slit and creates a dot on the photographic plate just behind the open slit. The same thing happens when that slit is closed and the other is opened. Now, what would the photon do when both slits are open? The intuitive mind would tell a person that, because there is only one photon, the photon would simply pass through one slit or the other. Whatever slit through which the photon traveled, one would reason, it would hit the photographic plate behind that slit, and create the dot. Each successive, independent photon would thus make a similar journey.

At first, that's what appears to happen. One by one, each photon moves through the slit and hits the plate, though not in the same spot, which in itself is a bit curious. Each of these photons hits on a different and apparently random place on the plate. But if one is patient and continues to watch, something even more curious and astonishing takes place on the plate. In time, as one photon after the other completes its journey, the observer eventually sees that an interference pattern is slowly forming on the plate. The separate, individual photons seemingly work together to create a perfectly formed interference pattern.[1] What's more, the bands of light and dark are detailed and precise—there is no blurring, no stray markings. Not one photon lands in a place that would disrupt the pattern. The photons each move to the exact spot needed on the screen to help in the creation of the pattern.

This raises several questions. First, if the photons depart from the same source, why do they travel a different ultimate path on their way to the plate? And second: how, as individual photons, do they recognize the situation—in this case, two slits being open as opposed to just one being open—and know how to act accordingly? To highlight this second question, it's important to remember that if one of the slits is closed off at any point, the interference pattern will vanish, only to be replaced by one dot (or beam) behind the one open slit, just as it would have if only one slit had been open to begin with.[2] The light will readjust to the situation of one slit suddenly being open, and thus will live out that outcome. This brings us back to the first of the above questions. Why do these photons travel different paths when leaving from the same source, with the same trajectory, speed, etc ...? If only one slit is open, their trajectory appears the same; they travel through the one open slit and hit the plate in such a way to form one beam of light. But when two slits are open, and a choice is given, their trajectory, and thus their behavior, appears altered by this circumstance.[4] One possible explanation is that maybe the slits are considerably larger than the photons themselves, thus giving the photons more room to maneuver when traveling through them. One would reason that if the size of the slits were decreased, so that there was only enough room for a photon to pass through when aimed and sent directly toward the slit, then the path of the photon would be reduced to a straight-line trajectory. In this way, one could better guarantee that the photon would move through the slit in a straight line, thus striking the waiting plate with far less variation. But this, strangely, is not what happens.[3] Instead, the opposite occurs. The path of the photon becomes more unpredictable, and the interference pattern is still formed, but spatially *larger* than before. It's as if by reducing the slit, the particle responds to the confinement and, in a sense, rebels against it. A similar experiment could be designed with similar results using other particles. For example, Fred Alan Wolfe, in his book *Taking the Quantum Leap*, describes an experiment in which a pencil is broken down and heated in an oven until reduced to a steam of atoms. These atoms boil out of a small hole in the oven and move toward a black screen that also has a small hole in its center. Behind this first screen is a second, white screen that contains an emulsion that can record the behavior of the atoms. Many of the atoms are caught on the first screen, but a few do make it through the tiny hole to travel and land on the second screen. What is seen is the same thing that was witnessed in the dou-

---

4. This leads to an interesting third question—a question that will not be dealt with in this essay. If each individual photon's path is altered during its *lone* journey in this variation of the experiment, then what's interfering with it? If the photon is solitary as it travels, then how (or what?) is interfering with it to change its course? This should only happen when multiple photons are present. This has provided further evidence to some, such as David Deutsch, that parallel universes exist and interact with one another.

ble-slit experiment—if one watches long enough, these apparently random atoms will slowly form perfect circular rings of light and dark on the second screen. Just like the photons in the double-slit experiment create bands of light and dark that resemble the slits, the atoms in this experiment create perfect circular rings. There is no blurring of the rings—but perfect circles of light and dark … light and dark. As Wolfe states, they look "like ripples created by a dropped pebble in a still lake". Once again, we see individual particles take on wave properties when placed in a certain situation.

At this point, Wolfe asks: would making the hole smaller in the first screen cause the rings produced on the second screen to become smaller, if not cause them to dissipate altogether? As in the double-slit experiment, just the opposite resulted. When the hole is cut smaller, the atoms respond in the same rebellious way by making even larger, but still perfectly defined rings on the second screen. And strangely enough, if the hole is cut larger, the atoms create smaller rings. Just the opposite of what the intuitive mind would expect. The question is: why? The answer seems to lie in Nature's reluctance to be observed or measured, particularly at the quantum realm. The closer the observation—which in the Wolfe example is the first screen's hole being made smaller—the more unpredictable, and often rebellious the reaction.

For additional illustrations of this phenomenon, we can again return to the double-slit experiment. To better observe the plight of the particles in the various scenarios, scientists have placed sensors on the slits—sensors which are able to examine the photons as they journey through the slits. What has been discovered is that when the sensors are in place, the particles react to the observation by changing their immediate actions and behavior.[4] When the sensors are removed, the particles instantly return to their previous behavior. When the detectors are in place, for instance, when both slits are open, the photons, as if recognizing the presence of the observing apparatus, exhibit particle properties. But when the detectors are removed, they again exhibit wave properties.[5] It seems that by trying to determine its position in a more precise manner, we changed a particle's trajectory, a consequence that falls right in line with one of the bedrock principles in the world of physics—the Heisenberg uncertainty principle. The uncertainty principle states that it is not possible to know both the position and the velocity of an object at the same time, because to observe one is to affect the other, thus making an accurate reading of both virtually impossible. This explains, for example, why physicists still don't have a full understanding of the atom. When attempts are made to determine the exact location of an electron in any given orbital, the act of observing it changes it, so that all one is left with is essentially a probability. We know where the electron *should* be, but we're not quite sure that's where it

---

5.   When the detectors are in place, the results coincide with human expectations—as if nature attempts to reveal to humankind only that which we are capable of understanding.

will be when we observe it. This also explains why electrons seemingly display claustrophobic tendencies when cornered. For instance, if an electron is trapped in some sort of adjustable container in an attempt to better determine its exact location, the electron, as the container is shrinking around it to observe it more closely, will become absolutely frenetic, ricocheting off the approaching walls with incredible speed.[5] The microscopic world apparently does not permit its components to become entrapped, and thus closely observed; when they are, their motion becomes unrestrained.

When examining the behavior of particles, in this case photons, in all of the aforementioned experiments, one of the central questions arises for those in the scientific community: are we witnessing examples of Nature, at its most fundamental level, displaying a discerning intelligence? When particles seemingly recognize and respond to changing circumstances; when they are apparently aware and reactive to observation; when they, as *individual* particles, actively conspire to create perfectly formed patterns when placed in specific situations—are they not, in all of these cases, displaying a pronounced level of discernment? If so, can those working from the strict scientific frame of reference continue to ignore the very real possibility that a force of intelligence may be the orchestrating principle behind it all?

At this point, an empiricist may still find it possible to disregard the behavior of light, in and of itself, as sufficient evidence to move the discussion forward. One could argue that the behavior of light could be just an isolated phenomenon in nature that will soon be explained away by some new theory based on the strict orthodoxy of materialism. "Intelligence" may not prove to be relevant in the final analysis. In the face of this argument, one would need to counter with other examples, other evidence, offered to us by the very underpinnings of physical existence that would seem to indicate that "intelligence" is a prime component in the workings of Nature.

Rudy looked up with a shake of the head, handing the paper back to James. "What does it all seem to be telling us?" he said as he again began moving along the trail.

"That's the exact question Dad always asked. 'What is the natural world showing us?' That's where he believed our focus should be, spiritual and otherwise."

◆   ◆   ◆

At just that moment, the brothers came upon a break in the woods. From where they stood, they could see the small parking lot through the bare tree limbs, which at this early morning hour only contained three cars, one of which belonged to David, Sinclair's uncle. James and Rudy made their way silently down the path, and as they did, Sinclair pushed out of the car's passenger door with his backpack in hand. Rudy was once again amazed at how big his son had become—long and lanky, like the rest of the men in the family. Rudy made his way over to the car, greeted and thanked his brother-in-law for dropping Sinclair off. When David's car pulled away, Rudy looked at his tired son. His brown eyes were a little puffy from an obvious lack of sleep.

"You and your cousin were up all night, huh?"

The boy shrugged. "Pretty much, yeah."

James walked over to where they were standing.

"Hey, Uncle James," Sinclair greeted him tiredly.

"How's it going, Sinclair?"

"Alright."

"Did you get something to eat?" Rudy asked his son.

"We stopped on the way."

"So you're ready to get going then?" Rudy asked.

Sinclair nodded. "I'll wake up soon."

Rudy laughed, put his arm around his son's shoulder as the three headed back onto the same path Rudy and James had just been traveling.

"How's school, Sinclair?" James asked.

"Okay. Seventh grade's been pretty easy so far."

Rudy walked ahead as he listened to Sinclair and James talking. They hadn't seen each other since the memorial service—and James and Sinclair had always been close. After about ten minutes, they found themselves back in a densely wooded area. Fortunately, the path they were on was well traveled and thus clearly defined.

"What a day, huh, Sinclair?"

"It's a lot warmer than it's been," the boy answered. The sun's pretty bright."

"Speaking of light, your Uncle James and I have been talking about the double-slit experiment."

"You and I just talked about that last week," Sinclair remembered.

"Pretty wild, huh?" James said.

"Yeah …" Sinclair answered, "it's kinda crazy."

"In fact," Rudy added, "we just finished talking about the Heisenberg Uncertainty Principle. You remember we talked about that, too, a while back?"

Sinclair nodded, trying to recall. "It was something about knowing a particle's speed and position … something like that."

"Not being able to measure the velocity and position at the same time," James reminded him, "because measuring one causes the other to change."

"Oh yeah," Sinclair nodded.

At just that moment, the woods came to another small opening, and the three men found themselves looking out over a small, briskly moving stream. Carefully, they made their way down the small embankment to the water's edge. They watched the moving water for several minutes before James broke the silence.

"Several things come to mind regarding what we've been discussing, Rudy," he said quietly. "First of all, the fact that these things seriously chip away at the notion of a smooth, continuous deterministic universe. With the observer affecting reality, and with the subsequent reality based in probability ... I mean, the causality of classical physics seems to break down at the quantum level. And second, as I alluded to a few minutes ago, if one examines the behavior of particles in all of the aforementioned experiments and scenarios, an intriguing question arises: are we witnessing examples of Nature, at its most fundamental level, displaying a discerning intelligence? When particles seemingly recognize and respond to changing circumstances; when they are apparently aware and reactive to observation; when they, as *individual* particles, actively conspire to create perfectly formed patterns when placed in specific situations—are they not, in all of these cases, displaying a pronounced level of discernment? If so, can we ignore the very real possibility that a force of intelligence may be the orchestrating principle behind it all?"

Kneeling down, Rudy laid his fingertips just inside the cold, clean water, and felt its brisk movement. "But some strict scientists wouldn't concede the point that there is an intelligence working behind it all. To them, such a proposition sounds too much like one is putting forth the notion that there is an external God manning the controls."

"And some in the religious community," James countered, "would see it in just the opposite way—as in fact being full evidence of the workings of God. This intelligence they would say *is* God." He knelt down. "In either case, it certainly seems that there's an intelligence working within Nature—a component of Nature. I mean, there's innumerable examples that seem to indicate this ... from Pauli's exclusion principle, to nonlocality, to the workings of valence electrons. As we've seen already with just the wave-particle duality of light, when not observed, the wave aspect dictates, but when observed, the particle property instantly takes over."

"A phenomenon which, by the way, is often referred to as collapsing the wave function," Rudy added.

"Okay," James continued, "so in essence, the observation gives the quantum system a definite position."

"You remember Dad talking about Schrodinger's cat?" Rudy asked.

James thought back. "I do vaguely."

"Schrodinger's cat?" Sinclair asked.

"Yeah," Rudy looked at his son, "we haven't talked about that yet. See, Erwin Schrodinger was an Austrian physicist who, in the 1930's, proposed a hypothetical and somewhat morbid scenario to represent the notion that the observer is responsible for collapsing the wave function, and thus determining the reality for any given system."

"That's right," James remembered, "didn't it have something to do with placing a cat inside some sort of chamber?"

Rudy nodded. "A closed steel chamber with radioactive material that has a fifty percent chance of decaying in sixty minutes. At the end of the sixty minutes, if it does decay, a circuit will be activated and the cat will be electrocuted."

"Yes, a bit morbid," James agreed.

Rudy pulled his fingers out of the water, shook them dry. "Let's be glad it's only hypothetical. But the question posed is: can the cat be considered alive or dead before the chamber is opened and the animal is observed to be one or the other?"

"The answer seems obvious," James answered without hesitation, "if the material decays, the cat would be dead—if it doesn't decay, it would be alive, even before the observation is made."

"In our macro world, yes, but not so according to the rules in the quantum realm ... which brings forth an interesting paradox."

"That's right," James recalled, "at the quantum level, the cat is neither dead nor alive, but both until an observation is made and the wave function is collapsed."

Rudy nodded. "As illustrated in the double-slit experiment, a wave can be in several places at one time, but when it is observed, it reacts to the observation, reverting to one position, essentially the particle. At this time, its wave function is collapsed. Before that, *as a wave*, it was in many positions at one time, and those positions were based merely on probability. Within a wave, there's a certain percentage of probability that the photon will be in one position, and a certain probability that the photon will be in another position. Its actual location is determined by the observation of it, and the subsequent collapsing of the wave function at that exact place and time."

"And this surely holds true for any quantum system," James offered, "even for the one found in the cat's steel chamber ..."

Rudy tapped his staff on a nearby rock for emphasis. "And all components of the system—the cat, the radioactive material, the inner walls of the chamber—are subjugated to the rules of quantum mechanics." He paused, stood back up. "Speaking of interesting questions, and a question that goes along with what you have proposed regarding the role of intelligence ... what about: what is the transition point between the micro realm and the macro realm? Where exactly *is* this transition point? Where's the boundary between the two?"

James nodded. "In a way you're asking if a bridge exists between physics and metaphysics, and if so, where?"

"It could be framed that way," Rudy agreed. "A significant portion of the answer can surely be gleaned from the crossover point between the micro and macro realms," then he paused, obviously working through a sudden thought. "And to take it full circle, it is ultimately, in a way, addressing the question currently dominating the world of physics—is there a point of unification between quantum physics and general relativity?"

"Einstein's unified field theory," James smiled, as if reading his thoughts.

Rudy nodded. "But before we move on to that, let's deal a bit more with the whole intelligence question."

James nodded in agreement.

"You mentioned Wolfgang Pauli's exclusion principle. Interestingly enough, it was Schrodinger who came up with a mathematical model for the atom that described electrons as waves ..."

James turned to Sinclair. "See, Schrodinger's model was based upon assigning a probability to best determine an electron's position within a certain region around the nucleus. These regions, or orbitals, are regions around the nucleus where the probability of finding an electron is the greatest." James paused. "I wouldn't imagine you've had any of this in school yet ..."

"Not yet," Rudy answered, "but we've talked some about it."

"See, Sinclair," James continued, "within the atom, there are principal energy levels, which in turn are divided into sublevels. The first principal energy level has one sublevel, the second principal energy level has two sublevels, etc ... It is in each of these sublevels where the orbitals for the electrons are located, and only a certain number of electrons can occupy any one orbital."

"And the occupation of an orbital by an electron is based upon something known as *spin*," Rudy continued. "Each electron appears to be spinning on an axis, and it can only spin in one of two directions—clockwise, also known as spin-up, or counterclockwise, known as spin-down. To occupy the same orbital, electrons must have opposite spins, thereby canceling each other out to obtain a total spin of zero. This is Pauli's Exclusion Principle."

He paused, looked at James. "You mentioned earlier non-locality, or what's also called quantum entanglement. This falls in perfectly with that."

"Quantum entanglement?" Sinclair asked.

James nodded. "If a previously interacting pair of particles, such as electrons, are separated and then independently observed or measured, the observation imposed upon one of the particles will *instantaneously* affect the other particle, regardless of the distance separating them. In essence, one electron could be located in the far reaches of the Andromeda Galaxy and the other in Rancho Palos Verdes, California, and when an observation is imposed upon one, the other will respond instantly—almost as if of one mind. If one particle contained a certain spin, for example, the other particle would respond with the corresponding spin instantaneously."

"And *instantaneously* is the keyword," Rudy added, "Einstein referred to this as 'spooky action at a distance'. This, along with other aspects of quantum mechanics, really troubled him."

"Because this notion of instantaneous signaling that quantum entanglement puts forth," James explained, "violates special relativity, which states that nothing, not even signals, can travel faster than the speed of light."

"In fact," Rudy added, "Einstein explored this further through the EPR experiment, which is named after himself, Boris Podolsky, and Nathan Rosen, all of whom were from Princeton. They found that the spin of the particles involved was not only preserved, which means that zero total spin was maintained, but this was accomplished because the spin of the second particle was actually determined by the outcome of the measurement of the first. Let's remember that when not observed, the particle is a wave, and its properties, not only its position but also its spin, is determined when the observation is made and the wave function is collapsed."

"And remember further," James reminded him, "the outcome is undetermined—it is based on probability. For example, there's a fifty percent chance that the spin will be spin-up and a fifty percent chance that it will be spin-down. So if the observation is made that collapses the first particle into a position where it possesses, for example, a spin-up position, the second, *regardless* of the distance separating the two, will instantaneously have a spin-down to maintain zero total spin. Now some, such as Einstein, argued that the spin was really determined at the time of particle creation."

"But didn't ..." Rudy snapped his fingers in an attempt to recall the name.

"John Bell?" James assisted.

"That's it! Didn't Bell, several years after Einstein's 1955 death, prove through a host of additional experiments that the spin was indeed determined at the *moment of the observation*. If the wave function is collapsed to indicate the first particle has a spin-down, then the second particle will respond instantaneously with spin-up."

Rudy shrugged. "The question is: spooky action at a distance, or one discerning mind at work?"

"Either way, certainly not static," James concluded.

"Or so it seems ..."

"Yes," James nodded in second thought, "or so it seems."

3

March (Afternoon)

James remembered being very young the first time the thought had occurred to him. He believed he was maybe four years old. It was summertime and he was sitting outside on the back lawn one early morning, looking up at the large tree that sat at the center of their back yard. He studied the fresh leaves and noticed how the sunlight seemed to darken them to an almost majestic shade of rich green. He remembered taking in a deep breath, suddenly wondering if all he was witnessing was only in existence because he was there. He recalled the exact questions he had, though in four-year-old language, asked himself: does this tree exist at this moment solely so I can experience it? Does the grass beneath me exist for the same reason? And the sky? And the dog barking in the distance? And what's more, will they continue to be here when I leave?

From then on, James had these same thoughts repeat themselves on numerous occasions. When he reached age seven or so, he remembered telling an older boy in the neighborhood what he had been thinking. The boy, who was about age fourteen or so, looked at him scornfully. "That's stupid," he said. "My mom would say that the world doesn't revolve around you."

So bothered was James by this response that he finally took the issue up with his father one late spring afternoon. "Hmmm," his father said, looking down at him from the top step of the front porch. "That's an interesting perspective, James." James then told his father about the boy's rebuke, and was surprised when his father laughed.

"Try to remember, James," he said gently, "that there's more than one way to see things. Some people only see the world one way and that's it." He then called Rudy outside. "Would you boys like to go out to lunch?" he asked. "Let's go over to that restaurant by the water."

"Can we get pizza, Dad?" Rudy asked.

He smiled. "I believe they have pretty good pizza there."

45

A few minutes later, their father maneuvered the car into the portion of the restaurant's parking lot that overlooked a vast expanse of the Severn River. He ushered them both out of the car to stand on one of the docks stretching out over the placid river.

"I want each of you to look out upon the water," he instructed, "and think about three things you notice. I'll give you ten seconds." James remembered looking out over the busy river, seeing a number of boats drifting and/or speeding past, the large houses on the far embankment, the seagulls flying overhead, and the children swimming just off the far shore.

"Well?" his father asked after a few moments had passed.

Rudy spoke first. "I saw the dark blue boat that said *Cincinnati* on the side, the seagull that flew down into the water over by the end of this dock, and ..." he pointed across the river, "the people rowing in the little boat way over there."

"And you, James?"

"I saw the three seagulls sitting on the pole near the far dock, the boys swimming and yelling, and the ripple in the water in front of us ... I think it might have been a fish."

"Whose description is right?" his father asked. "Whose vision of this river is more accurate, more valid?"

After a few moments of silence, Rudy finally spoke up. "Both."

"Why do you say both?"

"Because if I look now, I see what James saw."

"And James?" his father looked down at him. "What do you see if you look again. "Do you see the same things this time?"

James looked out over the river, and saw many more things, including what Rudy had seen. "I see all sorts of things. Even what Rudy saw."

"Do either of you see the bird's nest in the tree over by the water's edge?" He pointed to a tree near the restaurant's back entrance. Sure enough, in the highest limb of the tree was a thick nest made of leaves and a vast array of sticks.

Both boys shook their heads. "I didn't see it Dad," Rudy admitted.

"Does that mean that it didn't exist?"

They both looked at him.

"Do we *really* know for sure if it did or didn't?" he asked.

James and Rudy looked at each other. "It was there before I looked," Rudy said slowly. "It had to be."

"But how do you absolutely know for sure?"

"Well ... because you saw it."

"But do I really know for sure it was there before I saw it ... or when nobody at all is looking at it?" Then he smiled when he noticed their perplexed stares. "A tree falling in the forest, boys ..." Then he drapes his arms over their shoulders. "Someday you'll understand fully what I'm asking you." Then he looked down at James. "But one thing's for sure, James—your thoughts on the things around you in the world, though it may seem self-centered and even uncomfortable to your friend and his mom, is a valid perspective." He smiled at them reassuringly. "It reminds me of an old quote by a guy named Andre Gide—'one can only discover new lands by first consenting to lose sight of the shore for a very long time'. It takes courage to think outside what's commonly accepted, and even more courage to express it to others."

Rudy and James had looked at each other. This was one of those conversations that their father had engaged them in that neither son would fully understand until years later. With this particular conversation, James distinctly remembered Rudy asking his teacher the next day at school two things: what the word "consent" meant, and if he knew anything about trees falling in the forest.

◆   ◆   ◆

"All of this talk about the role of the observer in determining physical reality has me thinking," James looked over at Rudy as the three began to make their way along the stream, "on two fronts actually. First, the fact that observation/ measurement impacts physical reality, and second—that the physical world, at least at the quantum level, seems to respond in a discerning way."

"The sheer fact that the observer cannot be taken out of the equation regarding how the physical world responds and seemingly structures itself," Rudy agreed, "leaves many doors open. One of which, at least to me, is a renewed sense of hope."

"Hope? In what way?"

"Well, unlike the mechanical, deterministic view of the world we discussed earlier, this view leaves room for a continually active/reactive creative element in the structure of the physical world ... a creative element that allows us not to be completely excluded from the shaping of the reality we experience ..."

"Okay," James halted him, "I can see two reactions to that, one from the scientific community, the other from the religious community."

"And they would be?"

"From the science side," James explained as they again began to move slowly along the stream, "many would say that there is no substantial evidence to suggest that human observation affects anything, and thus is not a creative agent in regard to the structure of reality."

"My immediate response to that is yes," Rudy answered, "in the macroworld, no it does not appear so, though there are some cases, such as Dr. Masaru Emoto's experiments with water, that opens the question up a bit."

"That's right," James recalled, "wasn't he the researcher from Japan that used water from the Fujiwara Dam to see if mental stimuli could impact the molecular structure of the water?"

"It was. If you'll remember, in one case he had a Buddhist Monk bless the water, and indeed the molecular shape of the water appeared to change in pronounced ways. Other recorded cases dealt with the changing of the water's molecules after receiving mental stimuli that offered thoughts based in love, in hate, in appreciation, and the like."

"I remember now. And didn't the molecular structure apparently change each and every time in such a way as to reflect the intention being put forward? An offering of love caused the molecular structure to take on a beautiful shape and crystalline color that resembled almost that of a microscopically examined snowflake. The offering of hate, on the other hand, changed the molecular structure into a seemingly discolored, non-uniformed structure."

"Dr. Emoto asserted that the level of purity of the intention being sent forth influenced the water more than any other factor. He also contended that a pure intention, such as that found in meditation or prayer, could even affect the water from a great distance."

"But either way, even if findings such as Dr. Emoto's are completely discounted, and *human* observation proves not to, in any quantitatively verifiable way, affect the physical world in a direct manner, it's really kind of a moot point in that it doesn't change the fact that Nature indeed responds to observation/measurement at its most fundamental level. And what's more, we don't *really* know for sure if human observation, at some level, doesn't in fact impact reality. We just may not be able to detect it yet. I mean how many times has science itself made predictions before having the measuring apparatuses for verification of the prediction? The neutrino, for instance, comes to mind. To essentially preserve the law of the conservation of energy, Wolfgang Pauli predicted the existence of a particle that had no charge and zero mass and that, in short, held and moved unaccounted-for energy. He called this particle the neutrino, which means 'little neutral one'. He made this prediction in 1931, but it wasn't until 1956 that such a particle was verified to actually exist."

"What's the law of the conservation of energy?" Sinclair wondered.

"It states that energy is not created nor destroyed, but only changes form."

"That's right," James remembered, "when it was realized that atoms were losing more energy than they should have when electrons were emitted, the fear was that energy was being lost, destroyed, if you will, which threatened the law of the conservation of energy."

Rudy nodded. "So Pauli, convinced that the conservation of energy was correct and could not be violated, concluded and indeed predicted that there had to be a particle that we weren't aware of that was carrying away this missing energy. And it was roughly twenty-five years before he was proven correct."

"And then there's Einstein's 1916 general relativity prediction that space-time warps whenever a mass is present," James reminded him, "and it warps in the direction of the mass. But it wasn't until the 1919 solar eclipse, when starlight, not normally visible, was observed bending toward the sun as it moved past it."

"Exactly," Rudy nodded. "The overriding point is: even if human observation doesn't ultimately wind up impacting reality, the fact that observation, be it a detector on a slit recording information about passing photons, or a light source examining electrons in an atom, causes the physical world to react is, for me, not diminished in the least. The possibilities this holds for reality being in a continual state of creation remains a kind of Holy Grail in my mind. And the fact that Nature not only reacts, but apparently reacts in a discerning manner, opens up so many possibilities for what the nature of existence might be. It actually brings forth a host of spiritual-type implications."

"You know, everything you're saying would cause strict scientists to balk," James warned.

"I realize that, but it nonetheless doesn't change the fact that whenever an observer's in the mix, reality is affected … whether it's Einstein's special relativity, or the orbiting location of an electron in an atom, the observer plays a part in the outcome." He shrugged in resignation. "Until science can remove the role of observation from the total picture of reality, it will remain the wild card … and remain one of the essential and beautiful mysteries of the nature of existence."

"And it's a wildcard that could," James pressed on, "if one chose to do so, bolster a metaphysical, i.e. spiritual, take on the nature of existence."

"Absolutely," Rudy nodded, "which brings us to your second group of people that tend to balk at the significance of such things—some of those entrenched in the religious community, particularly the strict doctrinal segment of the religious community."

"In many ways, these phenomenal aspects of the physical world we are discussing give the religious community some quantifiable ammunition in their arguments for the impact of "intelligence" as a force in the structure of existence. As we've seen, the evolvement of scientific capabilities is beginning to show how Nature not only responds to occurrences and situations, but apparently does so in an intelligent, discerning way. This would seem to bolster the notion of an intelligent force working within it all."

"And many in the religious community do reference these things," James pointed out, "but they do so mostly with the take that they are tools utilized by God to structure his world. This in and of itself is not a problem, not until they take that next step of trying to place what Nature shows them within the framework of whatever doctrinal belief system they proclaim to be the truth. Unfortunately for many, the pure study of Nature, and what that study might lead to, has little meaning. Its wings are clipped so it can stay grounded within the story-based religious paradigm."

Rudy paused, leaned his weight onto his staff and stared down at the slowly moving water. "And they also tend to revert back to what always seems to be the crux of story-based religious belief—a human-centered framework. This is why the first question often asked is: does human observational capacity have any real input on the physical world? And as we've just discussed, this question misses the point entirely. Being a remnant of transcendent theology, it's the kind of question that has soaked into human consciousness over several millennia ... the kind of question that symbolizes our need for instant, and human-centered spiritual gratification."

"I'm not following, Dad," Sinclair spoke up suddenly, his eyes peering through his camera's view finder as he focused in on a rock formation further up the stream.

"See Sinclair," Rudy walked over to his son and leaned behind him to catch a glimpse of what his camera was focusing in on. "Basically what I'm saying is that many people will dismiss this entire conversation if we can't show an impact that our own observational capacity has on the world. The fact that we, in this human form, appear to not have the capacity to impact the macro world will be the death-knell for this conversation and all it entails in the minds of many individuals. So what, they will say, that some subatomic particles respond to a measurement/observation in some experiment in the way that we've described? What does any of this have to do with us, with our truth, with our destinies, spiritual or otherwise? This kind of sentiment, in the large majority of cases, is a direct result of century upon century of a belief system that has put forth the notion that for a vision of reality, a vision of truth, to be viable it must somehow place us, humankind, at the apex of the process, and as a central feature in the grand design."

"You know, made in God's image and the like ..." James clarified.

"As a result," Rudy continued, "if we're not a part of the central focus, we quickly fail to see the relevance, indeed the connections between one thing and the next. Suddenly there's separation between us and everything else. The quantum realm, viewed through this lens, is wholly separate and distinct from our reality, from our destiny. From this vantage point, the quantum realm, and indeed Nature itself, is impersonal, without spirit, and void of the substance of human spiritual truth."

"When in fact," James picked up, "this view is actually contrary to the popular vision of an all-encompassing God that many of these same individuals proclaim to hold."

"See," Rudy continued the thought, "there's that sense of human entitlement that exists within the popular 'established' spiritual community ... a sense of entitlement that actually swims against its own current in that it doesn't allow for their own professed belief in a full, all-encompassing God to be a reality."

"Whattya mean?" Sinclair asked.

"Well," Rudy thought for a moment, "God is, it seems, only all-encompassing when he is understandable and germane to the human element. It's as if God would never express himself, or provide a *profound* clue to his workings in something as seemingly impersonal and detached from the human condition as an electron or a photon."

James smiled at the irony. "Even though these things are *fundamental* in God's physical structure of the very world we depend upon for our survival."

◆   ◆   ◆

About ten minutes later, they came to a place where a large tree had fallen across a narrow portion in the stream, its thick trunk forming a bridge to the other side. Rudy studied it for a moment, and then looked over at his brother. "You wanna give it a try?"

James looked at the shallow stream flowing beneath it. "It's only about a foot deep. Yeah, let's go across." One after the other, they balanced their way across the sturdy trunk to the other side, and continued in the same direction along the water's edge.

Over the course of the next few minutes, James and Sinclair began talking about everything from school to the latest in music. At one point, Sinclair pulled the camera he had received from his grandfather on his last birthday from his backpack, and showed it to James. Rudy had told James on the ride to the park earlier that morning about Sinclair's sudden interest in photography. It was only a few days after his grandfather had passed away that Sinclair dug the camera, still new in its box, out from under his bed. Rudy believed it was Sinclair's way of continuing to bond with his grandfather—it was his grandfather, after all, who had bought him the camera and encouraged him, as was written in the birthday card, to "capture what is otherwise not seen." Rudy wasn't quite sure if Sinclair understood just what he had meant by this, but it sure seemed that he had spent the last few weeks trying to figure it out. As a result, the camera had been his constant companion.

Rudy watched as his son stopped suddenly, and motioned with his hand for them to pause and remain quite. He pointed to the break in the tree cover just ahead. "I swear I just saw a hawk land on one of those high limbs," he whispered as he knelt down onto one knee, and moved the camera into position. Sure enough, just a moment later, a large hawk took flight from one of the highest limbs, its magnificent wings stretching out like a beacon in the blue sky, and it glided seemingly without effort, it's call echoing—*"key-ahh … key-ahh …"*. Sinclair's finger went to work, snapping a litany of pictures before the bird drifted behind a nearby incline in the forest. Sinclair immediately looked at his small screen, and smiled. "Got it …!"

He held the camera up so the two men could see what he considered to be the best shot. "You sure did …" James nodded in approval. "What a beautiful animal."

"I think it's a red-shouldered hawk," Sinclair told them. "I did some research last week … I told my science teacher about our hike, and he said to look for hawks. He said right now is the beginning of their mating season. So I went on the internet the other night and looked at a bunch of photographs and stuff …"

"And this one," Rudy asked, "looks like … what kind of hawk did you say?"

"A red-shouldered hawk … I think."

Sinclair turned the camera around, studied the photograph. "Pretty cool … I read on the internet that hawks have eyesight something like eight times more powerful than we do."

"That's not surprising," James concluded as they balanced their way across a small rock formation. "From the heights they soar they need that keen vision to spot their prey. What do they hunt, Sinclair?"

"Small animals … sometimes even rabbits and squirrels."

"They must have some mighty strong claws to lift an animal the size of a rabbit off the ground," Rudy said.

"And good depth perception to be able to spot and then catch these types of animals while flying at such high speeds," James added.

Rudy shook his head as he stepped down from the final rock and dug his staff into a dirt path that now had formed along the stream's edge. "No wonder that rabbits spend so much time in underground burrows ... and are largely nocturnal."

"And when they are out during daylight hours they spend a lot of time in bushes and other covered places," James reminded him. "But even then, they have so many other enemies, such as raccoons and wolves, that being outside at all puts them in constant danger."

"A wild rabbit's lifespan is something like a year," Sinclair told them as he paused to place his camera in a small case and drape it over his neck.

"It's still interesting to notice the habits and physical makeup of animals such as a rabbit or a hawk, and see how they have been shaped by Nature," Rudy offered. "I mean, just the rabbit itself, with how it lives and behaves in its territory ... living underground and being most active at night. It's all an adaptation to its environment." He turned back to his son. "Don't they have an entire system of burrows that have many different entrances and exits?"

"I think my teacher called them *warrens*," Sinclair answered, remembering back to the last day in class.

"And don't rabbits also have this nasty little habit of occasionally spreading a different kind of dropping ... softer droppings that they actually eat?"

Sinclair nodded with a grin. "They're usually dropped in the early mornings ... my teacher says as an extra source of nutrition."

"Since they spend a large amount of daylight hours underground," James reasoned, "this makes perfect since."

"That's pretty astonishing," Rudy said. "I mean if you think about it. That the body of a nocturnal, vulnerable animal like the rabbit would produce such a self-sustaining system to provide an extra source of nutrition ... it's the body's natural reaction to the possibility that the animal may have to spend extended periods of time underground without access to food."

"And we see many cases where Nature has recognized similar needs in various animals and thus evolved adaptive qualities," James added. "Look at hibernation. Squirrels and groundhogs, for example, during extended periods of hibernation have the ability to reduce their heart rates from somewhere around 150 beats per minute down to less than ten. And they even reduce their breathing to the point where they only take a breath once every few minutes."

"Which cuts their need for the oxygen supplied in their blood," Rudy added, "thus enabling them to survive long periods of fasting."

"Like with the rabbit droppings," James continued, "they're all instances of Nature displaying a kind of situational awareness—recognizing the circumstances of a particular animal's reality and developing into it added safeguards. And examples of these adaptive qualities are all over the natural world—snakes living in areas occupied by large mammals developing rattles to keep from being trampled, the horse losing its outer toes in favor of a single hoof when it moved millions of years ago from a forest dwelling animal to an animal occupying open plains, and thus needing to move at high speeds to escape predators ..."

"Light colored moths in England being replaced with dark colored moths," Rudy added.

"What's that about?" Sinclair asked.

"Prior to the industrial revolution," Rudy answered, "peppered moths in England were primarily a light shade, and thus they blended in with light-colored tree trunks to avoid being detected by predators. But when industrialization took over, the soot from the factories darkened many of the trees, leading to the emergence of darker-colored peppered moths."

"However," James pointed out, "don't some people argue that these darker-colored moths may have already existed, so a whole new moth, if you will, didn't actually evolve in response to the industrial revolution, but just more of the darker-colored moths were now able to live and reproduce in greater numbers?"

"True ... and there could be some validity in their argument. But it still doesn't change what we're talking about here, which is the fact that a reaction took place within Nature, whether it was the creation of a new darker peppered moth, or the reemergence of the already existing darker moth. Either way, a response to the environment took place."

"And we see this in humans," James emphasized. "The Moken people, for example, off the coast of Thailand and Burma who live almost entirely on the sea. Because they live such aquatic lives, they have developed the ability to see underwater with double the clarity as that of the average human."

"And they can swim underwater for a much longer period than any of us can."

"And both of these abilities they have right from childhood."

"People who have lived in mountainous regions for generations are another example," Rudy continued, "such as those in areas of South America. They find their offspring being born with thicker, stronger legs."

"And they have a larger lung capacity to deal with the higher altitudes," James added.

"The lighter-skinned people of the world have ancestors that originated from colder climates, whereas the darker-skinned people's ancestors originated from the hotter regions. Both adapted qualities to deal with the different climates."

"Nature certainly shapes us in our macro realm, that's for sure."

"Everything's seemingly wired together …" Rudy concluded.

◆   ◆   ◆

As they maneuvered their way along the stream, James recognized how the direction of their talk was winding its way toward Einstein's unified field theory. He stretched his memory to recall all he knew about Einstein's pursuit of an ultimate theory. He remembered learning how the man had spent the last thirty or so years of his life pursuing the notion that the four fundamental forces of Nature—electromagnetism (light), gravity, the strong and weak nuclear forces—could be unified under one mathematical equation. He was especially concerned with finding a point of unification for light and gravity, which appeared to be two completely unrelated forces. At the time he was pursuing his theory, the large majority of the physics community was moving in the apparently opposite direction—that of quantum mechanics. What they were finding at the quantum realm appeared to contradict much of classical physics, and thus made the notion of finding a unifying theory of the kind Einstein professed seem even more unattainable. Though a grand unifying theory was still held up by most physicists as a kind of grand ideal, the majority began to abandon the search for it. Then several years after Einstein's 1955 death, several physicists were working on something then known as string theory, which was an attempt to explain the strong force. It was during this research that they stumbled upon evidence that the theory predicted the existence of particles representing both light and gravity. It was then they realized they might have just found the key to Einstein's last great vision. This theory eventually became known as the superstring theory.

"Rudy," he called out to his brother, who was walking several paces ahead, "what were the names of the physicists who originally worked on string theory—the ones who first stumbled upon the fact that light and gravity were apparently represented within the theory?"

"Are you referring to the guys from Cal Tech?"

"I believe so."

"But remember," Rudy reminded him, "there were two teams of physicists who came across light and gravity in the theory, but it was the second team from Cal Tech, led by a guy named John Swartz, who really chased it down."

"String theory?" Sinclair, who had paused to change his lens, looked up at them questioningly.

"I've talked to you about string theory before, haven't I?" Rudy asked in surprise.

"It's been a while …" Sinclair answered.

James laughed. "He must have been tuning you out, Rudy …"

Rudy smiled. "It's understandable sometimes, I suppose."

"Isn't it about vibrations and energy, or something …?" Sinclair recalled.

"Along those lines," James answered. "Albert Einstein believed that the four forces of Nature—light, gravity, the strong and weak nuclear forces—could be unified, something that no one really thought could ever be done."

"What are the strong and weak forces?" the boy asked.

"Essentially, in short," his father answered, "the strong force is responsible for holding the nuclei of atoms together, and the weak force handles radioactive decay."

"And these four fundamental forces are seemingly incompatible with one another," James clarified, "especially light and gravity, but Einstein believed that a point of unification could be found among all of them."

"When Einstein died in the 1950's, he did so without proving his ultimate theory, but nearly two decades later a team of physicists were working on a theory that was designed to explain the strong force when they realized the theory itself predicted the existence of photons, the quantum representative of light, and gravitons, the quantum representative of gravity."

"Two of the fundamental forces …"

"And the two that seemed the furthest removed from each other."

"This caused the theory to be abandoned by this first team of physicists, but a few years later Swartz and his team at Cal Tech picked it back up and it soon became recognized as the possible path to Einstein's unified field theory. The old man, they realized, may have been right all along."

"It eventually became known as the superstring theory, or string theory for short."

"Doesn't string theory predict the existence of something like eleven dimensions?" James asked.

"It does."

"So, if correct, we're truly living in more of a multidimensional world than we ever imagined?"

"It would seem so."

"What about those vibrations I remember you telling me about?" Sinclair wondered.

"That's essentially what the superstring theory says occurs. The theory states that all forms of matter are not point particles, as was once believed, but are in fact made up of microscopic vibrating strings."

"Billions of times smaller than a proton," James added.

"Essentially," Rudy continued, "according to the theory, the underpinnings of our world consist of these vibrating strings, which oscillate, or vibrate, at different frequencies to create the different forces and particles found in our world, *including* gravity and electromagnetism. Each vibrating mode of the string produces a particle whose characteristics are determined according to the string's particular oscillating pattern."

"Think of guitar strings, Sinclair" James suggested, "how the strings resonate at different frequencies to create a variety of notes."

"So these strings of energy vibrate one way to make one kind of particle," Sinclair attempted to clarify, "and vibrates a different way to make some other particle ...?"

"Essentially yes ... and the strings vibrate differently still to make the various forces, from light to gravity, found in Nature."

"So everything is just different vibrations of these strings of energy?" Sinclair concluded.

"If the theory's correct, that pretty much sums it up," Rudy confirmed. "*If* the theory's correct, and there's still a good bit of debate about that among scientists." he pointed out. "Though," he quickly added, "most of these same scientists, on both sides of the string theory debate, agree that there must be a theory of unification out there, be it string theory or some other theory yet to be placed on the table ... after all, the harmony of the natural world seems to clearly indicate that all of its components are interconnected, and scientists almost universally acknowledge this."

They walked in silence for the next several minutes before James spoke up. "The real question is: what does this all mean?

"Whattya mean?" Rudy wondered.

"If everything is unified, what does it tell us about reality, about our lives and our world?"

Rudy paused, pointed his staff toward a large tree that had fallen up ahead of them. "Let's take a rest for a few minutes," he suggested.

A few minutes later, the three sat on the thick trunk of a fallen hickory, sipping water and contemplating the notion of a unified world.

"Some religious thought," Rudy began, "particularly Eastern thought, has been saying for centuries that all things are truly interconnected."

"Some sects of Buddhism," James nodded, "talk about the notion of no separation, no distinction between one thing and the next."

"This is such a difficult concept for sentient beings to grasp, let alone structure their lives around."

"No doubt. Especially since all that we experience in a direct manner, in the ways that impact us the most as individuals, seem to tell us just the opposite. Our sensory experiences seem to tell us that there is definitely a separate self, and though all things are interconnected, they seem to be so in an indirect way. What I mean is that we recognize that all living things depend upon their environment for survival and all things play a part in the overall maintenance of the conditions for life, but we still believe that there is ultimately a separate self that experiences it all and does so independently of all else. After all, the pain I feel when I place my hand on the hot stove is my pain, independent of anyone else. It is exclusive to me. It's difficult for us to accept that there might be, at the most fundamental level, not a *me* that is truly separate and distinct. It goes against everything we have been conditioned to believe, particularly in the context of Western thought."

"But if the superstring theory, or some later theory, is found to be correct and all things are *proven* to be unified at the most fundamental level, we will have no choice but face this reality. The ultimate question is: how can we reconcile this knowledge with what we experience?"

Rudy took a long drink of his water, stared up through the bare tree limbs of the forest and noticed how clear the sky had become. "What a spectacular sky ... crystal blue." Then he looked at his brother, and took another long drink. "I suppose it's the same challenge that has faced those striving for spiritual awareness, spiritual connectedness all throughout history. The core of true spiritual striving is rooted in this same basic challenge—the sacrificing of the immediate self for a greater connectedness, a greater awareness, of whatever one deems to be the creative force behind their existence, whether it be "God" in the traditional western sense, or Brahman, or the Sioux's Wakan ... whatever the case may be."

"So this knowledge of the unified field being the structure of our physical reality, if indeed it is proven to be true, will force us to confront this same challenge?"

Rudy nodded. "I believe so, especially if it is someday proven through quantitative verification and evidence that people can tangibly see ... at that point, no longer is this challenge wholly abstract. Nature, through testing and experimentation, will have been unveiled. After all, describing a painting and letting people actually see it with their own eyes are two different things. They can believe that the painting is a work of beauty through the description they have been given by a third party and the faith they have in that description, but they will only actually know it once they can see it for themselves. The secret doubt many have regarding matters of faith and the subsequent belief it requires will be removed if the superstring theory, or some other theory of unification, is proven …"

"And thus the obstacles standing in our way when it comes to truly facing this eternal challenge will have been lifted," James added. "We would finally see the truth before us … clear and unobstructed."

Rudy stared back up through the tree limbs and marveled once more at the deep blue that seemed to encase them. "And a new stage in our journey as a species will have finally begun in earnest."

4

March (Late Afternoon)

Over the course of the next hour, the three pushed their way back toward the river. The temperature was now beginning to feel more like a normal March day. James, following behind Rudy and Sinclair, wrapped his scarf tighter around his neck and leaned forward against a now stiffening breeze. They made their way up a small hillside, the tall hickory and ash trees spreading out before them in all directions. Even in the cold weather, James loved the allure of the woods. A long hike in the isolation of wooded terrain brought a calming to his soul in a way that was unlike anything else. In contrast to the rest of his busy life, he felt not only the obvious connection to Nature itself, but a heightened connection to himself as well. The time spent in the woods, more than any place else he could think of, forced him to move outside of himself and thus, in some strange way, move closer to his own essence.

As they pressed up the hill, James' mind shuffled through everything that had been discussed throughout the day. The possibilities that the natural world was unveiling were truly amazing. Though some would argue that many of the theories currently being tossed about would ultimately not prove to be an accurate reflection of how Nature operated and how reality was truly structured, this refutation was not cogent from James' perspective. It didn't matter that some of the theories, such as the superstring theory may not, in the end, prove to be correct. What mattered to James were the unfolding possibilities they represented. Even if the current theories were not correct, James had little doubt that whatever theories evolved to fill the void would be just as seemingly unbelievable, mysterious and awe inspiring. To James, what modern investigation, via more advanced measuring apparatuses, was currently showing might not be the actual and final answers to questions regarding the true structure and origins of the nature of existence, but what they were opening up in the human mind were the seemingly unimaginable possibilities that existed. This, to James, was the underlying point, and all that ultimately mattered.

The discussion of unification, whether via the superstring theory or through some other means, fascinated James. If it was proven in a truly quantitative way that all things were in fact unified at the most fundamental level, and it was no longer a matter of debate, what impact would that really have on humankind's overall approach regarding life and existence? James figured that, in some ways, components of the perennial science/religion debate would continue unabated, though the focus would surely shift almost entirely to the question of the origins of unification. Where did this energy originate? Was an intelligent force responsible for it? Was this method of unification, and for that matter, unification itself, just the way that God (whichever God one believed in who was making the statement) chose to structure his world? Or, as many rationalists would probably continue to maintain, did unification occur solely through chance?

Here again, to James, the fact that many aspects of the rationalism versus dogmatism debate would remain intact did not diminish the significance of a new, indisputable knowledge that all was indeed unified. This knowledge would narrow the debate, push it forward, and realign the focus. In this way, Rudy made a good point—if unification was proven, a new stage in the human journey would surely begin.

Angling his upper torso to a point where his shoulders were almost parallel to the ground, James pressed further up the hill's gradually increasing incline, his staff now bearing a considerable amount of his weight.

"Are your legs burning, little brother?" Rudy called back to him.

"Lungs too," James answered.

Rudy laughed, but he too was beginning to labor. Luckily, it was only another forty or so feet before the hill leveled off and they found themselves moving down a slight incline, the flow of the river again filling in the background.

Continuing to reflect on the day's discussion, James felt a gnawing realization that there were many people who would have found the entire exercise futile. Some would say that nothing was solved, while others would say that it was all just fanciful, impractical, and speculative musings. An understanding of ultimate reality, they might say, can never be attained by humankind anyway, so what's the use? Then there were those who would feel that an understanding of ultimate reality might be attainable, but his and Rudy's discussion offered nothing substantive—after all, virtually no conclusions had been drawn and no answers had been put forward. It was just disconnected bits of information that had little to do with piecing together an ultimate vision of reality, let alone the reality the average person experiences each and every day.

As their pace settled into a more relaxed rhythm, James let his mind circle carefully over this notion of ultimate reality. To him, "ultimate reality" and the "nature of existence" were, as far as terminology was concerned, one and the same, and he recognized that there were many interpretations of what they both might mean. Some might believe that ultimate reality equate to a direct knowledge of "God", while to some it might be an understanding of the strictly physical workings of the universe. What's more, there might be some that view ultimate reality as being the state when one becomes in tune with whatever principal awareness exists that is responsible for structuring the reality of human experience. And then there were those who believe that an ultimate reality simply does not exist. Whatever the case, James was struck that, regardless of the particular interpretation each individual held, they were all searching, in the end, for the same thing, even the individual who asserts that an ultimate reality does not exist. The searching was all directed to the one great set of universal questions: why am I here? where did I come from? where am I going?

It occurred to James that what defines each individual is not the motivation behind the inquiry into these questions, but the method chosen to search the possibilities. How do we each go about arriving at our version of "truth"? Is one means more valid, more accurate than the next? As James pondered this, a pair of questions equally fundamental kept appearing and reappearing in his mind: what are we capable of knowing? What are we designed to know?

For James, the answer to these questions rested, first and foremost, with our physical relationship to the world. Nature—or God, for those working from a religious frame of reference—had bestowed upon humankind five senses, all limited in their scope, with which to experience the world. The restricted capacity of our five senses brings up an immediate question: if our senses are indeed limited, how can we ever come to truly *know* anything approaching an ultimate reality? After all, there would always be more to reality than we would be capable of experiencing, so how could we ever hope to arrive at truth through any means purely dependant upon sense perception? And what's more, our senses can sometimes be fooled. James wondered if it was a similar line of thinking that led many people to dismiss scientific pursuit as the primary means to know truth. In fact, James had often encountered a kind of disdain for science among a surprisingly large number of the people he had known. Many expressed the sentiment that science often seemed completely blind, sometimes arrogantly so, to the more ineffable nature of life and experience. James found this line of thinking to be, in itself, a problem of perception, a problem of perception fueled internally within some segments of the scientific community itself, and a problem of perception externally imposed upon science by those working from a completely different paradigm altogether, most notably those of a strictly metaphysical bent.

James often thought that the problem of perception within the world of science was one that would require constant soul-searching by each scientist regarding their private motivations for why they devoted their lives to science in the first place. James guessed that many were guided toward scientific pursuit because they were enticed—even *moved*—by a version of the same basic questions a spiritual thinker might have been moved: why am I here? where did I come from? where am I going? Was it the case, James wondered, that some in the scientific community had become almost prisoners of strict empiricism, to the point where the beauty, the mystery and the possibilities of the unknown had begun to lose their allure? Had all possibilities for an unknown that was also unexplainable ceased to exist for the strict rationalist? Was there no longer the possibility of something existing that could not be explained through scientific methodologies? If so, could one indeed then argue that the ineffable possibilities of life were no longer of value to someone entrenched in such a paradigm? Was losing sight of this, James pondered, the great danger that lurked within the world of the scientist? After all, beauty and love aren't experimentally quantifiable, but they are essential in weaving the underlying fabric that creates meaning and fulfillment in the human experience. It seemed to James then that for science to be wholly viable, it was essential for it to maintain a view toward seeking a truth greater than itself.

With this said, James thought about the dismissal of science by some of the people he had known. The theories put forth by science were often wrong, they would say, or at the very least constantly evolving and changing. How then can the one truth that surely exists for the great questions ever be found in such a system? What's more, they would argue, science often gives virtually no final, definitive answers to the questions of truth—it's dependant solely upon cold fact and evidence, and thus tells us only bits and pieces about ourselves and about any greater truth that might exist. Furthermore, the meanings of experiment, of measurement can be skewed by interpretation of the individual scientists involved in structuring them. They could design measurements and experiments in such a way that the outcome would reflect their own ideas and positions. Finally, how do we know that the tools of science are truly capable of showing us what's really happening? After all, look at how many times science has been mistaken and/or misguided. To the large majority of these people, science as a whole was far too uncertain as a means to know truth; for them, the truth would be found within the structure of religious-based thought.

Maybe it was just him, but the direct criticisms of science often expressed baffled James. The argument that science wasn't valid as a means to know truth because it was often wrong, was constantly changing, and was too uncertain were expressed as if these components of scientific pursuit were failings. To James, the search for truth was an ongoing process, and thus needed to be a living exercise—one that needed to adapt and adjust, to grow and change. After all, for a species with God/Nature-given sensory limitations that needed information about the world to fall within the range of five senses to be known, an investigation that actively engaged in seeking out and translating information in a way that was compatible with our senses would be imperative. Such an investigation would seem to be natural to our species, and would seem to be compatible with how Nature/God designed us. It doesn't seem that we were naturally structured, whether by "God", or by whatever force of creation one may believe was responsible for our existence, to be able to understand the world in an immediate, complete way. Science seems to be a perfect reflection of this natural reality. We were not meant to know truth in one great swoop, James mused, but the thought of knowing it all at once, without enduring painful and frightening periods of uncertainty does, however, reflect the need for immediate gratification that tends to plague the human animal. And this, one might present as a counter argument to the critics of science, is indeed the true failing in our search for truth.

The three made their way to the bottom of the incline. Rudy pointed his staff due east. "The car's over about a quarter mile."

"Where are we heading after we leave the park, Dad?" Sinclair asked.

"We have one stop to make before we head home."

"Where?"

"Where your grandfather grew up," Rudy answered.

Sinclair nodded, and the three again moved in silence. As they approached the car a few minutes later, they paused at the edge of the woods just at the opening of the parking lot. James looked at Rudy. "If you listen closely," he said, "you can still hear the river."

Rudy nodded and, with a slight grin, glanced at his son.

Sinclair smiled crookedly. "I know, Dad. I know ..."

◆   ◆   ◆

Sinclair stared out of the back window as they maneuvered out of the park lot and onto the highway. Under the setting sun, he watched the trees along the side of the highway flicker past.

"Why are we going to grandpa's old house?"

"The woods near his childhood home is where he wanted his ashes spread."

Sinclair leaned forward. "Are you gonna do that today?"

Rudy kept his eyes on the road ahead of him as James stared off through the passenger side window.

"That was our intention," Rudy answered quietly, then he looked through the rearview mirror at Sinclair. "You know, he requested that you be there too."

Sinclair leaned back. "You think we actually will?"

James shrugged. "Dad said to do it whenever we felt we were ready."

"What did he mean?" Sinclair asked.

James shrugged. "He didn't say. He only said we'd know when we arrived there."

Sinclair thought for a moment before asking the inevitable. "Are you ready?"

Both men were quiet for a moment. "I don't know, son."

Sinclair leaned up again. "Uncle James. Could I see your urn?"

James nodded, and removed the silver chain from around his neck, handed it to the boy.

Sinclair studied what he had always thought, up until just a few days earlier, was a simple pendant with a figure eight carved into its side. "You know, I had no idea until you told me the other day that this was not just a piece of jewelry, but was actually a little urn that held grandpa's ashes."

"Nobody knows it," James answered. "I wear it all the time, and people have no idea."

"Could I see yours, Dad?"

While keeping one hand on the steering wheel, Rudy carefully removed the chain from around his neck and handed it to his son.

Sinclair studied the two identical urns. "That's so wild that grandpa's ashes are in these." Then he traced his fingertip over the figure eight. "Infinity, huh …? He leaned back, dangled the two chains in front of him. "Did he really believe in infinity?"

James nodded. "He didn't believe that there was an ending point, only some kind of a continually changing and evolving state of consciousness."

Sinclair placed both chains around his neck, stared back out the window. "It's not time yet, Dad," he said quietly."

Neither man answered, but both knew he was right.

5

April (Morning)

One behind the other, the group of five moved up the Tobacco Ridge Trail. "We should be coming to a great view of Douthat Lake anytime," Kevin called to Rudy from the back of the procession.

"Gotcha," Rudy called back, his voice laboring a bit. They had been at it for more than forty minutes without a rest.

"How's everyone holding up?" James asked from his position behind Sinclair and Rudy.

"So far so good," Nancy spoke up from behind him.

"And you Sinclair?" James asked his nephew, who was only a few strides in front of him, the long tripod bag dangling off of his shoulder.

"Okay, Uncle James," the boy called behind him.

Though streams of sweat were dripping off his lips, James managed a smile. "You're a trooper, Sinclair," he said, and meant it. The three of them had left Maryland at 5:00 am to make the four hour trip down to Douthat State Park. They had planned their April hike just three days after the Patapsco State Park hike the previous month. They had contacted Kevin and Nancy, two old friends from Rudy's days in college, to see if they wanted to meet them. Kevin and Nancy were graduate students, and already dating, when Rudy was finishing his undergraduate degree nearly twenty years earlier. Rudy's dorm room just happened to be next to theirs, leading to the three becoming close friends. When the two were married several years later, Rudy was the best man. As luck would have it, after their marriage the couple moved to within just fifteen minutes of Rudy and James' hometown. As a result, they became close to the entire family, including James and Rudy's mother and father. After eight years of being practically members of the family, Nancy's job relocated and the couple, along with their two babies, moved to their current home just outside of Roanoke, Virginia.

A few minutes later, Rudy came to a break in the tree line, and suddenly found himself staring down at a magnificent view of the lake below.

Sinclair was the next to arrive. "Wow," the boy said as he leaned his weight onto his staff, staring out at the beautiful vista. "What a great view."

One after the other, the rest of the group arrived. "Breathtaking, isn't it?" Nancy asked.

"Sure is," James nodded, and then smiled, noticing Sinclair already pulling his camera from his backpack. "Can get some good pictures here, huh Sinclair?"

"Sure enough," the boy said as he fixed the camera's strap around his neck.

"How about we unwind here for a little while?" Kevin suggested.

"Good idea," Rudy nodded, swinging his backpack off his shoulder.

A few minutes later, the group sat alongside the trail, sipping from bottles of water.

"How far away is your home from the park?" James asked the couple.

Nancy and Kevin glanced at each other, both shrugging. "About a half-hour or so," Kevin said.

"How often do you make it over here for a hike?"

"Now that the kids haven gotten a bit older," Nancy explained, "we make it over three or four times a year. Usually in the spring and the early fall."

"The weather's actually ideal this time of year," Kevin said, marveling for a moment at the crisp blue sky.

"It's actually warmer than I thought it would be," Rudy observed.

"It's gotten warm early this year," Kevin agreed. "It feels more like the second week of May than the end of April. The rhododendron have already begun to bloom."

James looked around, noticing how the leaves on the trees were already a rich green. But it was indeed the rhododendron that caught the eye. The burst of pink, white and purple flower clusters seemed to be everywhere.

"Yeah, the weather for our hike last month was perfect too," Rudy was saying, "especially for the end of March."

"If I remember from our phone conversation," Kevin recalled, "you went out to Patapsco. What a nice park."

"Yeah it is," James agreed. "We had a nice hike."

"You know," Nancy said gently, "I think it's great you guys are trying to meet up every few weeks like you are. Your father would be touched."

Rudy nodded. "He loved the outdoors, and to hike."

James smiled. "He liked taking long walks and talking. He always said he did his clearest thinking in the woods."

"Your father had some pretty heady ideas," Kevin nodded, remembering some of the conversations he had with the man over the years.

James smiled. "We discussed during our last walk many of the things he used to talk to us about."

"Yeah, Rudy mentioned that during our conversation," Kevin nodded. "It's been a long time since all of us talked about some of those things."

Nancy handed her husband a second bottle of water. "I hear you guys discussed quite a bit, from Maxwell's propagation of light to the superstring theory." She smiled. "It's been quite a while since we've all delved into such topics together."

Kevin took a long drink. "What did you delve into exactly?"

"Well," James tightened the cap back onto his plastic water bottle, "as Nancy said, we talked a bit about Einstein's belief in a unified field. But we first found ourselves talking about the discontinuous and uncertain nature of quantum physics."

"And obviously the role of the measurer/observer came up quite a bit," Rudy added.

"Ah, the Copenhagen interpretation," Nancy offered with a smile.

James nodded. "I've done quite a bit of reading over this past month on some of the things we talked about during our last hike, and that happened to be one of them. Niels Bohr's principle of complementarity really presented a challenge to Einstein."

"What's the principle of contemp ...?" Sinclair asked.

"Contemplementarity," Nancy helped him.

"It basically says that a particle doesn't have a determined path or velocity until it is observed," James explained.

"In effect," Kevin added, "it doesn't exist until the observation of it is made."

James reached into his backpack and pulled out his iPhone. "Rudy and I found some writings on our father's computer. I emailed them to myself so I could access them through the internet."

"Ah, modern technology," Kevin exclaimed.

"Actually not so modern anymore," James reminded him as his fingers worked the tiny keyboard.

"True," Kevin admitted.

"We found these particular writings in a folder my father had entitled *Civics through Physics*. It must have been one of the projects he had been working on. Here's a segment where he was giving brief descriptions of some of the great scientists and their basic contributions. Here's what he wrote about Niels Bohr."

James squinted down at the little square screen and began reading:

Born: Copenhagen (1885–1962)

Contribution: *The Bohr Atom*

Regarding atoms, Bohr postulated that electrons exist in definitive regions at various distances from the nucleus. What's more, he postulated that the electrons revolve in orbits around the nucleus. He designed an atom that consisted of a nucleus being orbited by electrons, and these electrons had several possible energies which corresponded to several possible orbits, all of which were varying distances from the nucleus. He went on to conclude that an electron must be in a specific orbit (energy level), and that it cannot exist between orbits. It had already been determined that energy is not emitted in a continuous stream but in tiny discrete packets called *quanta*. Bohr reasoned that when a quanta was absorbed by an atom—for example, a hydrogen atom—the electron would "jump" to the next orbit (energy level), apparently not existing in the space in—between. In addition, when an electron descends to a lower orbit, it emits a quanta of energy in the form of light.

Contribution: *The Copenhagen View*

At the fifth Solvay conference (which was named after the Belgian industrialist Ernest Solvay) in 1927, the world saw the greatest physicists gather together to discuss and debate new ideas, particularly those dealing with the new quantum theory. Of all the great debates, the most heated and most fascinating, by far, were those between Albert Einstein and Niels Bohr surrounding the role of the observer/measurer upon "reality". Up until 1927, it was widely believed that the "external" universe existed entirely independent of the observer measuring it. Einstein and others believed that the universe indeed operated independently from the observer, and what's more, the universe was smooth and continuous, based exclusively upon the rules of causality. Niels Bohr, on the other hand, believed that the universe was in fact affected by the observer, and was, in the final analysis, not smooth, but discontinuous and acausal. This view prompted Einstein to utter the famous words: "God does not play dice with the universe." According to Bohr's Copenhagen view (also called the principle of complementarity), reality at the quantum level does not exist until it is perceived. A particle, for instance, is without a set position or velocity until an observation of it occurs. In effect, the observation dictates its state. Its existence, and how it exists, depends upon how we choose to observe it, and what we choose to observe about it. Bohr pointed to the wave/particle duality of light as an example. In short, photons (light) apparently behave differently in different situations. How we choose to set up our measuring apparatus will determine how photons will behave. In some situations, they will behave as a particle, and in some situations as a wave (see the double-slit experiment). This wave/particle duality, it turns out, holds true for all other particles as well. Bohr also pointed to the Heisenberg uncertainty principle as further evidence that the observer plays a role in the construction of physical reality.

Even with the reluctance shown by Albert Einstein and others from the "reality is continuous, causal, and independent from the observer" viewpoint, the Copenhagen View became, and has remained, the officially accepted view—a view which presents us with new and exciting possibilities concerning the nature of reality.

James paused. "And here's what it says in relation to Werner Heisenberg."

Born: Wurzburg, Germany (1901–1976)

Contribution: *The Heisenberg Uncertainty Principle*

At the time of Heisenberg's ascent, there was an active discussion (and some debate) going on about how objects such as electrons and photons (the quantum carrier of light) appear to exhibit both wave and particle properties. Sometimes they behave as waves, sometimes as particles ... all of which seem to depend upon the particular circumstance in which they find themselves. There was little debate, however, that a particle had a specific position at any one time. There was also little debate that a moving particle had a particular direction in which it was traveling at any one time. In 1927, Heisenberg proposed his *uncertainty principle*. The *uncertainty principle* states that it is not possible to measure or observe a particle's position and velocity at the same time, because by measuring or observing one is to change the other. You can know either the position or the velocity, but not both; the more of an exact determination you have on one means the less of an exact determination you will have on the other.

In classical physics leading up to Heisenberg, the equations that had been designed to predict motion depended upon knowing both the initial position and velocity of a particle. Now, the uncertainty principle tells us this is not possible. In fact, according to Heisenberg, you can only make approximate measurements of velocity and position. Heisenberg's conclusions placed in jeopardy the entire notion of causality, which was basic to all of human understanding of Nature and indeed the processes of life. The underpinnings of nature, of physical reality, thus seemed to be based, at least in part, upon randomness and probability. In essence, Heisenberg seemed to show that one could not observe Nature, at least at the quantum level, without disturbing it to some degree.

"Boy, that sure sounds like your dad," Kevin said with a smile. "I remember him talking to me about the Solvay Conference when I first moved to Maryland all those years ago."

"I tell ya, I would have loved to have been a fly on the wall at that conference," Nancy said with a shake of the head.

"Apparently Bohr and Einstein went at it pretty good," James offered.

"The whole Copenhagen view apparently really troubled Einstein," Rudy added. "From what I've read, he actually found it somewhat appalling ... that an observer could have such a continuously profound impact."

"Einstein was, for lack of a better term, a realist," Kevin pointed out. "He was from that scientific school of thought that believed that the physical world, the universe, existed and operated free from the involvement of an observer. The Copenhagen interpretation challenged this standpoint entirely."

"And it's amazing because Einstein was, in many respects, the father of quantum mechanics," Nancy explained. "He was, by all accounts, the individual who most saw the significance of Max Planck's discovery that ultimately indicated, at least to Einstein, that light energy was absorbed and emitted as discrete quanta."

"Einstein's work with the photoelectric effect placed back on the table that light was not just a wave, but also a particle ..." Kevin continued.

"But that opened up the entire can of worms—when is it a particle, and when is it a wave?" Rudy asked.

"This is indeed where the trouble began for Einstein," Nancy said, "as it sounds like you two discussed during your last hike, according to quantum mechanics, it depends upon the situation in which light finds itself. In certain situations it behaves as a wave, in others it behaves as a particle."

"Which draws the observer, the measurer into the equation," James extended on her thought. "Depending upon how it is observed, how the measurement is structured, is how light will behave."

"And when we observe and thus determine its velocity, we no longer know its position," Rudy added. "And vice versa."

"Thus the uncertainty of quantum mechanics that disturbed Einstein so," Nancy said as she opened a second bottle of water. "According to this new view, no longer was the structure of reality independent of the observer. What's more, Einstein believed wholeheartedly in causality. If event A happened, the laws of physics dictated that a certain set of specific responses would happen. The universe was thus smooth and continuous, just the way he intuitively believed it to be."

"And then along comes an observation and everything begins to jump all over the place," James added, "suddenly making the universe not smooth and continuous but jittery and disjointed. He didn't believe reality was ultimately structured this way."

"And it was this basic discussion that was the centerpiece of the Solvay Conference," Nancy reminded them. "The Einstein-Bohr debate over the role of the observer …"

"Over the next decade, Einstein put together a host of thought experiments, most notably the EPR experiment, in essence trying to prove that an observation of a particle's velocity and position could indeed be made at the same time."

"We talked about that last month," James said. "And wasn't it ultimately determined that Bohr's interpretation of reality seemed to be the correct interpretation?"

Nancy nodded. "By most physicists, yes. But Einstein continued to hold on to the belief that, at the very least, quantum theory was incomplete."

"Besides the acausality caused by the observer," Kevin interjected, "Einstein was also bothered by the presence of infinities."

James and Rudy glanced at each other.

"What?" Nancy asked, noticing that the brothers were looking at one another.

Rudy smiled. "Nothing—that's just where we began our discussion last month."

"With infinities?"

"Dad was fascinated with the prospect of an infinite universe," James clarified.

"Which Einstein himself believed in," Nancy pointed out, "but not, it seems, in infinities existing *within* the infinite universe."

"What troubled Einstein about infinities?" Rudy asked curiously.

"Well," Nancy shifted her body so she was sitting Indian style. "From what I understand, when an infinity occurs in a mathematical equation it means the equation has failed. So in the world of physics, where the language is mathematics, the occurrence of an infinity has always meant the end of a prospective theory, at least where classical physics has been concerned."

"But in this new physics," Kevin added, "infinities are popping up all over the place, even in general relativity. And Einstein believed that infinities could not be a feature of reality. For instance, at the center of black holes it is believed that singularities exist."

"Singularities are basically infinities," Nancy clarified. "And at the heart of black holes, space-time becomes so warped that neither space nor time continue to exist." "And density is infinite," Kevin explained. "And it is here that singularities occur … the point where classical physics completely breaks down."

"And in the Einstein school of thought," Nancy continued, "this signaled not a feature of reality, as some physicists maintained, but a flaw in quantum theory itself."

"And from what I've read," James added, "infinities also pop up in other places too, such as with the description of fields and the unending number of variables they contain."

Kevin nodded. "The basic point is, at least for someone like Einstein, for quantum theory to predict infinities meant that it was incomplete as a theory."

"But couldn't it be possible that infinities are in fact a feature of reality we are just not yet able to measure, and thus to fully understand?" Rudy wondered.

"Well, in some ways quantum physics seems to be conceding that this might just be the case."

"And this is where the rub is between scientists—the realists on one side and those who are willing to reach out beyond the strictly realist approach on the other."

"What if Einstein was wrong and Bohr was right," a voice suddenly added itself from above the sitting adults—a voice not yet heard in the discussion.

"Hey, another precinct heard from!" Kevin laughed.

Sinclair smiled down from a standing position behind his tripod. "Yeah, I've been listening," he said as he adjusted the height of the tripod in preparation for his next photograph. "So what if …?" he asked again.

"Well," Kevin stretched his legs out in front of him, shaking his head slowly as he thought through Sinclair's question. "As we've been seeing from some of the theories that have arisen out of quantum mechanics, from nonlocality to the possible existence of multiple universes, the implications are potentially incredible."

"I read a bit about superposition over these last few weeks as well," Rudy explained. "The fact that a piece of the subatomic world, such as an atom, can exist in two places at once is, in and of itself, amazing, but it becomes even more amazing when we realize that it will assume one of the locations the moment it is observed."

"That's right," Nancy confirmed, "an atom could be, for example, both above and below a closed spacecraft at the same time, and will remain in both locations as long as it is not interacted with, including being observed."

"But once an observer appears and attempts to locate the atom," Kevin completed her thought, "it will instantly choose one location or the other."

"What we talked about during our last hike," James reminded Rudy, "it will collapse at the moment of the observation."

"That's exactly right," Nancy nodded.

"We also talked a bit about quantum superposition, or quantum entanglement, last time too," Rudy told them.

"Another amazing feature of the quantum world ..." Kevin shook his head, "... that two formerly entangled particles seemingly continue to communicate with one another in an instantaneous fashion ..."

"Across vast distances ..." Nancy added.

James nodded. "During the Patapsco hike, we asked: spooky action at a distance or one discerning mind at work?"

"Hmmm ...," Nancy sat back, propping her weight onto her elbows. "Interesting question. What do you mean when you say *mind*?" she asked.

James looked at her pensively for a moment, and then twisted off his bottle cap. "Good question," he said as he took a sip of water. "When I think of mind, I think of something that has always been in existence ... in existence, if you will, as the carrier of consciousness."

"So the mind is not something that originated within the brain, but is something that originated independently from it?" Nancy asked.

"That's how I've always thought of it."

"Is mind, to any degree, located within the brain?" she wondered.

"Each being has been given access to a portion of, what I've always referred to as the mind stream."

"A stream which itself flows from consciousness?" she pressed.

James nodded. "That's kind of the way I think of it."

Kevin pulled his knees up, wrapped his arms around them. "For me, the question is: what's your definition of consciousness?"

James took a long moment to think through his answer, and then tried to choose his words carefully. "To me, consciousness is the initial breath of life ... the first flash that signals existence, within the world and within each individual. It's like the electricity that sparks a light, and the light itself is the mind. Once it flickers on, it illuminates the world for all beings. To my way of thinking, consciousness is like the electricity that fuels the light ... it is the source of all existence, the substance that permeates it all."

"Let me clarify—when you say 'within the mind of each individual being', are you saying each individual has a separate and distinct 'mind'?" Kevin asked.

"No, not at all ..." James' fingers picked at the label on his water bottle, "like I said a moment ago, each individual possesses a piece of the greater mind. I've always envisioned a vast ocean with tributaries branching off of it. The ocean is the vast, all encompassing mind, and each being is a different tributary … and each is granted access to this vast ocean while we exist. And our individual brains are the inlets and ports we utilize in our everyday lives to gain access, limited though it may be, to the greater mind … again, what I call the mind stream."

"And consciousness is the substance which breathes life into everything that exists, including the mind itself." Rudy added.

"Breathes life into the individual mind," Nancy asked, "or, as you call it, the 'all-encompassing mind'?"

"Both …" James answered, "I've always believed, in the end, there's no distinction between them."

"So consciousness permeates it all?" Kevin clarified. "Not unlike the role of energy in string theory?"

"In a sense …" James thought about it. "If we play with the string theory as an analogy for a moment, the mind plays a similar role to that of the vibrations produced by the energy … And the individual mind is similar to a particular vibration that will ultimately produce a specific property of our world. The mind, like the vibration, is the portal of entry into the concrete experiences of our lives …"

"And the brain," Rudy followed up, "would be in a sense the actual mode of translation from the abstract into the concrete, not unlike the actual particles and forces themselves that are produced by the vibrations. The brain creates and builds the structure of life and existence that we experience."

"And just as the particles and forces are ultimately just various manifestations of an all-pervasive energy, the source from which it all arises," James added, "the mind and brain, and all they allow us to experience, are ultimately manifestations of consciousness."

"Someone might say that the string theory analogy is not an analogy at all," Kevin said with a slow shake of the head. "They might say that since everything in our physical world is made up of energy, then could energy and consciousness not in fact be one and the same?"

Nancy glanced at her husband, shrugged. "Then what is the origin of this energy, this consciousness?"

"Ah, an infinite regress …" Kevin answered with a smile. "It's like asking who created God?"

"Maybe a better way to frame the discussion," Rudy interjected, "is to ask a more preliminary, but still fundamental question. Assuming for a moment that consciousness and energy are in fact one and the same, do they possess intelligence?"

"Well, we talked a great deal already," James reminded them, "particularly during our last hike, how Nature, and thus the energy from which it is made, seems to possess not only intelligence, but a kind of discerning intelligence. I mean, when we look at the reactive nature of light, how it responds and adjusts itself to various situations, when we discussed the Pauli Exclusion Principle, and how electrons maintain zero spin within any given orbital ..."

"And the communicative nature exhibited during non-locality," Kevin added, "or the improvisational nature that particles exhibit in cases of superposition."

"And even something as fundamental as electron configuration," James pointed out, "which I also read about the other day in Dad's writings."

"Electron configuration?" Sinclair asked.

Nancy looked up at him. "I guess you're a bit young yet to have taken chemistry in school."

"Probably not for another year or two," Rudy nodded.

James pulled his iPhone from his backpack, and worked his fingers over the tiny keyboard for several moments until he found what he was looking for. "Here it is," he said as he began to read ...

When again looking at the structure of the atom, as well as the interaction between atoms, particularly in regard to the exchanging of valence electrons, we see clear indications of what seems to be an intelligence that recognizes and responds to various situations. For example, as electrons are filling the orbitals that surround the nucleus, they do so in a way that ensures that the lowest energy level available will be occupied first. An electron will not fill a higher energy level orbital before the lower energy orbital is filled. For example, lithium has an atomic number of three, which means that it contains two energy levels. The first energy level will be filled with two electrons, each with an opposite spin from the other, before the second energy level is filled with the third electron to complete the element's atomic structure. Those elements with a higher atomic number will naturally have more energy levels to be filled, and they will be filled in the same manner—from lowest to highest, with the lower energy levels being completely filled first before the occupation of the higher energy levels have begun. A very organized, precise procedure. The electrons occupying the outermost energy level, where the attraction to the nucleus is the least, are called valence electrons. These electrons are involved in the bonding of atoms—an interaction that serves to form compounds. With the exception of helium, the stable valence level is eight. In short, all atoms are constantly attempting to achieve, for themselves and others, a stable outer electron structure (which would make them a noble gas). To achieve this, they are in a continuous process of sharing electrons as needed. For example, a fluorine atom, which has nine electrons, seven of which are in its outer energy level, could pick up one electron from an atom of another element to complete its outer structure (in which case it would attain the electron structure of the noble gas, neon, which has ten electrons, eight of which are located in its outer level). Likewise, a sodium atom, which has eleven electrons, eight in the second level, and one dangling alone in the outer level, could transfer away the one lone electron to obtain a completely filled second level, which already has eight electrons. In this way, the sodium atom would have also obtained the same electron structure as that of neon.

If we examine further the interaction between two atoms, we can see the detailed and discerning process that unfolds in the formation of compounds. For instance, when one examines the electron structures of sodium and chlorine, they will see how the atoms of these two elements will work together to ensure that at least one or the other will attain the stable electron structure, and thus form the compound sodium chloride. Sodium has an atomic number of eleven, which means it has a total of eleven electrons. The eleven electrons are arranged so that there are two in the first energy level, eight in the second energy level, and only one in the third energy level. Contrast that with the chlorine atom, which has an atomic number of seventeen. Of its seventeen electrons, two are occupying the first energy level, eight are in the second energy level, and seven are in the third energy level. When sodium and chlorine atoms react with one another, the atoms recognize the other's configuration, and respond with an appropriate transfer of electrons so that one or the other obtains a stable structure. In this case, the one electron from sodium's third energy level will transfer to the third energy level of the chlorine atom, which has seven valence electrons, only one short of obtaining a stable structure. When this transfer takes place, the chlorine atom becomes stable, and the compound sodium chloride (NaCl) results. In essence, each atom has become an ion, held together by the resulting positive and negative charges. In this case, the chlorine atom, since it added the extra electron from the sodium atom, became a negatively charged ion (Cl-). On the other hand, the sodium atom, since it had lost an electron, became a positively charged ion (Na+). This now opposite charge possessed by each (through their recognition of the other and subsequent interaction) is what holds them together in what is called an ionic bond. The question is: purely responsive action/reaction, or discerning intelligence at work?

While James had been reading, Nancy had pulled a small pad of paper and pen from her backpack. Sinclair was now sitting next to her and she was drawing out for him what James was reading. When James stopped reading, she quietly explained it to him in further detail.

"Electron configuration is such a basic part of chemistry," Kevin said, "that how amazing it is often escapes us."

"Just one more example of an apparent intelligence acting out in the natural world," James said as he put away his iPhone.

"How about the consciousness side of the intelligence question?" Kevin wondered. "First of all, when we say that consciousness and energy may be one and the same, are we saying that they are one thing, or that they are manifestations of each other?"

"That's a great question," James admitted. "My first impulse is to say that energy is the fundamental force for the physical world and consciousness is the fundamental force for the abstract qualities of our world, such as thought, emotion and the like ..." "But thoughts and emotions are brought to fruition by a functioning brain," Kevin said, "which itself feeds off of energy."

"Thoughts themselves are the firing of chemical-electrical impulses within the brain," Rudy added to what Kevin was saying, "and these chemical-electrical impulses are themselves a form of energy. So consciousness, to exist, has to have energy, and energy is, in all the examples of intelligence in Nature we've been discussing, conscious."

"So what you're saying James," Kevin clarified, "is that there's a duality that exists ... the physical world of energy and the abstract world of consciousness, but at the same time, they're both a part of the other?"

"In a sense, yes. Let's look at it from a different scenario all together," James suggested. "Let's take a perfectly constructed robot—a robot that physically looks and feels human, has all perfectly simulated human organs, including a brain. The brain is programmed to think and to respond, even to emotional situations, by the same physical processes that an authentic human brain does ... even so far as to fire off chemical-electrical impulses when it 'thinks'. It is programmed to respond to almost any situation, and it even has a program that allows it to learn new information on its own, without the hand of its programmer to ever again be directly involved in its operation." James paused to ask the question he was leading up to. "Is energy being utilized?"

Kevin thought for a moment. "Obviously, it would be."

James nodded. "So the next question is: does this apparently fully functional robot—with its fully functional, malleable brain—have a mind?"

"Using your definition of mind," Nancy spoke up, her lesson to Sinclair now finished, "I would have to say no. The robot, at some point, had to have been purposely programmed for it to exist and to function in a way that simulated the human. But the mind, again in the way you framed it, only exists through natural processes ... as ultimately a manifestation of consciousness. Something can be physically and purposely created, even by humans, that utilizes energy, but something cannot be created to utilize mind if mind didn't already exist within it." She looked at James. "Do you agree?"

"I do," he nodded.

"Then you're saying that there is a duality that exists," she pointed out.

"A duality in the *manifestation* of each," he clarified, "like the flip side of the same coin."

"But you say that energy has consciousness, but in your robot scenario, you're saying that the robot does not have consciousness even though it is utilizing energy. Are you saying that energy is therefore not on the same par as consciousness? That they are, after all, not equal parts of that same whole?"

James shook his head. "On the purely physical realm, we can manipulate energy for practical purposes, but it is not energy in a pure, *undirected* state. I mean, we have learned to build cameras that utilize and manipulate energy to take photographs to recreate an image of an object, but we intuitively know that the photograph is in fact not the real object, only a representation of it. So I think that there's a distinction to be made between pure, undirected energy and the utilization and manipulation of this energy, just like there's a distinction between a photograph and the real object."

"Well then, can consciousness be manipulated as well?" Nancy pressed.

"Do our brains not freely use the part of mind we have been granted?" James countered with a question. "For example, are the individual perceptions we all hold, in the end, not manipulation of mind, unintentional though they may be?"

"And we're limited in this mind manipulation by our physical make-up, our limited senses," Rudy added, "just as we are limited in our manipulation of energy by our levels of technology and know-how."

"Again, we're back to the notion of a constantly evolving reality that we play an active role in creating," James concluded.

"We indeed are," Nancy agreed. "And as such, we probably should spend some time discussing that amazing apparatus between our ears that you seem to be alluding to as being a kind of passageway into mind."

"And ultimately into consciousness itself," Rudy added.

"But before we delve into the brain," Kevin interjected as he pushed to his feet, "whattya say we move onward toward the falls?"

"Sounds good," Rudy said, lifting himself off the ground with a groan.

James laughed. "Who's hurting today, old man?" he said as he slowly unfolded into a standing position.

Rudy returned the laugh as he watched James struggle to his feet. "Not much better, Junior."

"Just stiff," James said as he stretched his lower back.

"All of you are old," Sinclair said as he practically sprung off the ground, swinging his tripod bag across his shoulder.

"Don't ya love a show-off," James said with a shake of the head, pointing to his staff. "Now hand me my cane, Sonny."

6

Rudy—

Here's some more … In the heading, he referred to this writing as being "purely heuristic, but great fun nonetheless …"

James

Cc: Kevin and Nancy

… What about the seemingly pervasive role the observer plays in the determination of physical reality, as illustrated in the double-slit experiment, the EPR experiment, and even in the theory of special relativity? The apparent discerning responsiveness of the natural world at its most fundamental level to the act of observation brings us back to the individual ascending from Plato's cave. As posed earlier, the most intriguing question about the entire scenario may in fact be: what effect does the presence of this newly introduced observer have on the outside world? If the evidence presented in some of the aforementioned experiments and phenomena are accepted as correct, the act of observation ultimately does in fact determine and shape reality, thus lending to it a certain level of creative input. At this point, the natural question to be posed is: how does this all occur?

Some speculation … to keep the lifeblood flowing

An ideal launching point for this discussion may be with the age-old philosophical question: if a tree falls in the forest and no one is there to hear, does it make a sound? In essence, does the tree and the sound of its falling exist if there is not an observer present? The question seems to assume that the forest and all of its inhabitants do in fact exist without any debate (debate such as: how can they exist if not being observed themselves?), so we'll work from that premise. As written, one could argue that there might in fact be, right from the outset, a possible fundamental flaw in the question as it is originally posed. The possible flaw is contained in the phrase: "if *no one* is there to hear it". These words seem to imply that the "observer" needs to be of human origin to be valid in the outcome of this thought experiment. But is this so?

To begin to explore this in more depth, we would do well to turn first to James Clerk Maxwell's propagation of light, and then to Einstein's unified field theory. Though we have known about electricity and magnetism since the days of the ancient Greeks (and Chinese), it wasn't until the 1860's and James Clerk Maxwell that we became aware of the breadth of the relationship between the two. We had known that a moving magnetic field produced a moving electrical field, but it wasn't until Maxwell put together a series of equations that we came to know that the opposite was also true—that a moving electrical field produces a moving magnetic field, and that the two move across space and time, one producing the other, at the exact speed of light (electromagnetism). It was at this time that we realized that light was an electromagnetic wave. At this point, a question arises that relates to observation: if a moving electrical field creates a moving magnetic field, which subsequently creates a moving electrical field, and if they move across space-time as a wave, then what about thoughts that are produced through brain function? If thoughts are the firing of chemical-electrical impulses by the brain, are they not moving electrical fields? If so, do they create propagating fields akin to what Maxwell proposed? At the very least, wouldn't their fields impact other fields in the way that Michael Faraday set forth?

With this very heuristic proposition in mind, let's return again to our forest. If there is "no one" there, will the falling tree be heard? Would a better way to rephrase this be: if there's not a creature there that possesses brain function, will the falling tree be heard? What about the observation of one of the animals that surely inhabit these woods? Is their observation any less valid than a human observation? If we look again at the notion of the observer collapsing the wave function in any quantum system, be it in Schrodinger's steel chamber or the double-slit experiment, the act of observation itself must have a physical mechanism in which to make the observation. If it is a detector, then we would assume that the detector utilizes particles of some kind (such as photons) to make its observation. In essence, the detector shines light on whatever is to be observed. This light would be the physical mechanism of this observation device, and Nature at its fundamental level, as we've discussed, reacts in seemingly discerning ways to being observed. But what about pure observation, made without a man-made mechanism to actually perform the observation, such as that of an average human? If for the sake of discussion we accept the very tenable premise that human observation—be it prayer, meditation, or intense contemplation—does in fact have an impact on the reality we experience, then how is it done?

All throughout Einstein's theory of special relativity, there are countless examples of how the relative position, the frame of reference of the observer determines the reality at hand, and that all frames of reference are equally "correct" in their interpretation. We've heard of experiments, such as those conducted by Dr. Masaru Emoto, where, if true (for our purposes here, we'll assume that they are true), the molecular structure of water was apparently altered by mental stimuli. A pure intention/observation, such as that found in meditation/prayer, could even affect water from a great distance. The question is: what causes this? It can't just be the malleable nature of water in and of itself, but seems to be a demonstration of the ability of intense observation (which is what meditation/prayer basically is) to affect the physical world. As much as prayer/meditation is an odyssey from the brain into the mind and ultimately into consciousness itself, the operation of the brain function involved is the mechanism, the manifestation of human observation in the physical world. In the case of Dr. Emoto's studies, if they are indeed true, these chemical-electrical impulses seemed to be directed toward the obtainment of a certain aim, and apparently brought results. How can something like this occur? Why does observation, via prayer and/or meditation seemingly impact reality? It is here where it might be interesting to turn to Einstein's unified field theory.

As discussed earlier, Albert Einstein spent the last thirty years of his life pursuing what he called the unified field theory. He believed that the four fundamental forces of Nature—gravity, electromagnetism, the weak and strong nuclear forces—could be connected, or unified, within one mathematical equation. He was especially concerned with finding a point of unification for light and gravity, which appeared to be two completely unrelated forces. At the time Einstein was pursuing his theory, the large majority of the physics community was quickly moving in the apparently opposite direction—that of quantum mechanics. Though a grand unifying theory was held up by most physicists as a kind of grand ideal, many had long disregarded it as being unattainable, and somewhat unrealistic. However, several years after Einstein's death, *through* the study of quantum physics and what became known as the superstring theory, the possible key to Einstein's unified field theory (and the possible key to the fusing of general relativity and quantum mechanics), may have been unearthed.

In short, the superstring theory states that all forms of matter are not point particles, as was once believed, but are in fact made up of microscopic vibrating strings billions of times smaller than a proton. Essentially, the underpinnings of our world consist of these vibrating strings, which oscillate at different frequencies to create the different forces and particles found in our world, *including* gravity and electromagnetism. Each vibrating mode of the string produces a particle whose characteristics are determined according to the string's particular oscillating pattern. The superstring theory, if it is correct, has moved us one giant leap closer to Einstein's unifying theory.

Now, if we return to our discussion on the chemical-electrical impulses of the brain in relation to our understanding of Maxwell's equations, and filter in the prospect of Einstein's unified field theory, we might see an explanation as to how the chemical-electrical impulses of thought may cause at least a ripple within the fabric of space-time. This would also seem to fall in line with non-linear dynamics—better known as *chaos theory*—which is commonly understood by the statement that "one flap of a butterfly's wings affects the entire universe". Is this not implying that kinetic energy, in whatever form it takes, including chemical-electric impulses, affects the entire field?

As interesting as these speculations might be, they still do not answer how prayer/meditation, or intent of thought, such as that demonstrated in Emoto's experiments, can affect the "physical" world. To potentially solve this puzzle, one might be best served to approach the subject counter intuitively. We're searching for how the physical world is affected by brain function, so we tend to attempt to look at the outer manifestations of thought. But maybe the exact opposite approach should be examined.

A good starting point to do this might be with a discussion of a portion of the frontal region of the brain called the parietal lobe. The parietal lobe is the part of the brain that gives a person the feeling that they are separate and distinct; it is responsible for the feeling that a "me" exists, that there's a boundary between the individual and the rest of the universe. During deep meditation, prayer, or contemplation, the activity in this part of the brain is reduced, and thus the illusion of a "boundary" between a person and the rest of the universe is diminished. There is no longer the sense of separation. If energy may prove to be, in the end, the manifestation of consciousness itself and if, as the super-string theory proposes, all are vibrations of energy, then consciousness may in fact be the creative force of reality, and therefore all barriers, or sense of boundaries between any one thing and consciousness are indeed an illusion. When in deep meditation, prayer, or contemplation, one empties the "self" and merges with pure consciousness—what some call the *Clear*—and thus has access to the entire field of consciousness, the field upon which all exists. They are no longer confined within time and space, because space-time itself is only a vibration, and thus ultimately a construct of consciousness. It is here that what appears to be the direction of the prayer in an outward sense is in fact not an outward direction at all, but an inward journey that, once the Clear is reached, dissipates the illusion of a separate self. While in the Clear, there is no separation between this and that, between one and all. The outer manifestation of the meditative state (or the outcome of prayer) actually occurs at the moment one crosses this inner threshold—what this author calls the "event horizon"—and enters into the Clear. The prayer is originally directed outward in a purposeful direction, but at the moment the meditative event horizon is crossed, the *intent* of the prayer enters into and becomes pure consciousness, and thus, in an almost circular fashion, reenters the mind stream and gains access to all of existence, including the intended "receiver" of the prayer. It is at this point that the impact of the "prayer" upon the reality we experience is realized. This, in the end, is a process of pure observation, and meditation, prayer, and contemplation are the most intense forms of pure observation attainable by the human species. Such practices as watching (observing) the breath, examining (observing) a koan, reciting (observing) a mantra or prayer, and the intense contemplation (observation) of ideas and art leads one into a realm of creative possibilities that can impact reality—a reality we then experience as sentient beings. Through deep observation of this kind, we lose ourselves as we merge into vibration, and then into mind, and ultimately into pure consciousness, at which time there is no separation between "you" and the source itself, be it Brahman, Buddhahood, Allah, the Christian God, or Einstein's cosmic religious experience. Unification has been reached and "you" are now consciousness, and thus too a creator. It could be here, at this point of unification, where creative phenomena that we often interpret as being healing and divine revelation take place. And the more one enters this state, the more creative input in the structure of reality one will have. All great artists, writers, musicians, for example, ultimately delve into regions of the subconscious mind to find the final direction in the forging of their works. Many will tell of an epiphany late at night, or entering a working "zone" where the final inspiration appeared without force of effort, or of having a vision in the moment just before entering a deep sleep. For in these moments, the artist is no longer separate from his painting, or the musician from his composition, or the writer from his manuscript. The activity in the parietal lobe is reduced, the sense of "me" is gone and pure consciousness is experienced, if for only a fraction of a moment. And in this moment, the individual is merged with all that is, and a creator is born. In this moment, Muhammad hears the angel Gabriel, Jesus speaks to his Father, Buddha is awakened, and Einstein pens the final equation to the theory of relativity.

April (Mid-Morning)

The group moved along the Tobacco Ridge Trail, the bright sunshine gently growing in intensity as each minute passed. The trail wound downward for about a half mile or so until it branched off into the Blue Suck Falls Trail. Rudy marveled at the beautiful world of shadows the sun was creating along their way. As they pushed uphill on the dirt path, an endless array of forest shadows drifted over them, and there was an almost perfect quiet.

A few minutes later, they crossed a small stream that reminded Rudy of the Patapsco stream they traveled along during a good part of their last hike. From here, they continued on the main trail for about three quarters of a mile until their path descended into a sharp, rocky decline.

"We need to take it slow here," Kevin warned. "But the falls is just a few hundred yards ahead."

Carefully, the group negotiated their way down the steep, jagged path until they heard the sound of falling water. After another hundred feet or so, they moved into an even richer shadow world that surrounded a ledge about fifteen feet off the ground. It was from here that a stream of water fell into a small pool. "So this is the great beast, huh?" Rudy asked with a wry smile.

"This is it!" Kevin laughed as he and Nancy guided them to a large flat rock formation near the bottom of the falls. "We can rest here for a while," he said as they eased their backpacks off. Nancy laughed as she noticed Sinclair staring intently at the tiny waterfall. "Luckily we had a good two days of rain last week, otherwise this mighty waterfall wouldn't be much more than a trickle."

Kevin smiled. "It's not exactly Niagara, but it's still enchanting, hidden away here in the woods."

"It sure is," Rudy nodded as he took his backpack off and eased down on one of the large rocks. "Enchanting is indeed the right word."

A few minutes later, the entire group was sitting together, sipping water and enjoying the sound of falling water.

"It's amazing how calming this all is," James commented.

"It relaxes the mind, that's for sure," Nancy added.

"In fact, the mind's where we left off," Rudy reminded them.

"Indeed it was," Kevin nodded.

"I believe Nancy mentioned the brain," Rudy said, opening a bag of peanuts for the group to share.

"I think so," Nancy nodded, taking a long drink from her water bottle.

"When I think of the mysteries of the brain," Rudy continued, "the first thing that comes to mind is meditation, particularly in relation to the Tibetan Monks."

"Are you referring to the self-immolation of the monks in Vietnam during the 1960's?" she asked.

"Among other things," Rudy nodded. "But actually that was one of the first things I learned about regarding meditation ... and it indeed blew my mind. The fact that these monks could sit ablaze in meditation and not move or scream while their bodies burned to death is utterly astounding."

"And on the opposite side, you have Tum-mo," Nancy told him.

"What's Tum-mo?" Sinclair asked.

"A meditative practice that means 'inner heat'," she explained. "This practice is very prevalent among the Tibetan Monks."

"And it takes various forms," Kevin added. "The simplest consists of monks wrapping themselves in wet sheets in temperatures around 40 degrees and meditating ... meditating until their body temperature rises to the point that the sheets will literally start steaming."

"Thus the 'steaming monks'," James said with a nod.

"Correct," Kevin confirmed, "and what's amazing is that they suffer no ill effects, not even a shiver. Studies done by places like Harvard University have shown that they are able to manipulate their skin temperature upward some seventeen degrees."

"And obviously this shouldn't be able to be done, physiologically speaking," Nancy added. "A host of negative health effects should result, but nothing occurs."

"There's even been documented cases of these monks sitting outside in the Himalayas," Kevin continued, "where the snow is constantly falling and the temperature dips sometimes near zero. In these conditions, they meditate overnight wearing only light shawls." Kevin shook his head in amazement. "Again, they suffer no physical consequences—they simply rise up in the morning in perfect health. You or I would certainly die from hypothermia if placed in such conditions, but apparently these monks are able to manipulate their metabolisms in such a way that their bodies can handle these extreme conditions."

"The question is how?" Rudy said, staring back and forth between Nancy and Kevin.

"Well," Nancy adjusted herself so she was sitting with her feet tucked underneath her, "in the last email you sent us, your father spoke quite a bit about the parietal lobe. That's definitely a good place to begin."

"The reduction of parietal lobe activity would certainly seem to go a long way in explaining how a person could withstand physically painful situations," Kevin began, "I mean, if the sense of 'me' is reduced, then sensations attached to one's body would have to certainly be diminished as well."

"This makes sense," Rudy nodded. "When you think of those monks in Vietnam during the 1960's setting themselves on fire without any visible reaction, it's as if they were not even there ..."

"They may not have been," Nancy said, "I mean if the functioning of that area of the brain was reduced to such a degree that it no longer fired off impulses, then were those monks truly *there* as 'individuals'? Were they capable of experiencing anything as 'individuals', even sensory sensations?"

"Intriguing questions," Rudy said with a nod. "But just because that section of the brain was reduced to a practically nonfunctioning point, does that mean that the monk himself, in his *totality*, was no longer there?"

"Well if you accept the premise that this form and degree of contemplation can help one truly dissolve the self and merge with pure consciousness, then was only a shell of the monk left once the parietal lobe was reduced to a nonfunctioning level?"

"If the brain is what Nature has given us to experience and 'know' the outer world, but to also, even more importantly, serve as the conduit into consciousness itself, then is this what is not meant to happen? Have we not been given the potential tools within the brain to dissolve the self and merge, as your father said, into mind and then ultimately into consciousness? Has the brain been designed as our potential vehicle to do just this? If you accept this, then maybe those burning monks, besides their bodily shells, were in fact not present."

"So what you're saying, in a sense, is that they moved into another reality altogether?"

Nancy shrugged. "I mean, a person could frame this any way they like. A person of western faith could look at this as a point of transcendence into the presence of God. An eastern mystic might view this as a journey into a next incarnation. A strict scientist, meanwhile, might view this as simply an area of the brain no longer functioning and thus providing no sensation to the conscious individual, with the word 'conscious' being used in a more traditional context."

James pulled in a deep breath. "Maybe the ultimate question is: have our brains potentially been hardwired to experience ultimate reality?"

"And is it," Kevin added "possibly just a matter of our species evolving to our brain's full capacity?"

"But we have the entire problem of being limited by five senses," James reminded them.

Rudy shook his head. "I'm going to take a long shot here, guys," he said with a grin. "And let me just say up front that my level of understanding of what I'm about to say is at best minimal, but here goes: if the brain can restructure itself as meditation apparently seems to cause it to do, then wouldn't our senses, along the way, be restructured also? I mean, aren't our senses ultimately controlled and dictated to by our central nervous system? It only stands to reason then that if our brain begins to restructure itself in an evolutionary sense, then the senses have to, in the process, follow suit."

"And thus would no longer be a hindrance in experiencing the world at the 'higher', more evolved level," Kevin offered. "Interesting indeed ..." he said as he pushed his weight forward from his elbows, wrapped his arms around his knees. "And probably not that far-fetched."

"You know," Nancy added, "when we talk about meditation's impact on the brain, numerous studies have shown that high level meditation increases gamma wave activity ... and gamma waves are the brain waves associated with perception and consciousness. And what's more, I believe that it was a combined study through Yale, Harvard, and MIT which indicated that the areas of the brain that deal with attention and the interpreting of sensory impulses actually grow in those who meditate ... If I'm not mistaken, the study even revealed that areas of the brain that usually shrink, or narrow, actually grow ..."

"What you are really referring to is something I recently read about," Rudy interjected, "and that's neuroplasticity."

James nodded in agreement. "Which is an area of neuroscience that's been garnering a lot of attention lately, precisely because it sheds light on some of the things we've been talking about here ..."

"I haven't had a chance to read much on it," Kevin admitted. "What's it all about?"

Rudy tightened the cap to his water bottle as he began to explain. "Essentially it is the finding that the adult brain, in a process called neurogenesis, can generate new neurons, and thus rewire itself. It had long been accepted thought that the brain was structured and set for life in childhood ... but recent findings indicate that this may not be so."

"My teacher told us something about all of this," Sinclair, who had been devouring peanuts as he listened, spoke up. "He said that there are certain times, like ages two and around twelve that our brains can learn a lot, and if we don't use it then, we sort of miss our chance."

Rudy nodded. "What he may have been referring to is synaptic pruning."

Sinclair looked at him askance. "Okay ..."

Rudy laughed. "We need to take a moment to look at how the brain operates to understand it. The best place to start is with neurons, the nerve cells of the central nervous system responsible for sending and receiving electrochemical signals to and from the brain. Our brains have about 100 hundred billion neurons, and it has long been believed that something like ninety-five percent or more of them were formed by the age of five or so. But as we'll see, we're no longer entirely sure this is true. In any event, these neurons have incoming and outgoing extensions that look sort of like branches. These extensions are responsible for communicating information from one neuron to another. The outgoing extension, which sends information, is called the axon, and the incoming extension, which receives information, is called the dendrite. Both of these have many portals along the neuron, so multiple signals can be sent and received. And where the signal is received—the connection point, if you will—by one neuron from another along these dendrite and axon routes is called a synapse."

"So in other words," Nancy clarified, "when an axon is bringing a signal to a neuron, the point where it connects with that neuron's dendrite is called the synapse."

"Correct. And for each neuron there are upwards of 10,000 synapses."

"Once the signal reaches the synapse, what occurs?" Nancy wondered.

"In short, an information-laden chemical is released called a neurotransmitter, which connects to the cell membrane of the neuron at specific locations called receptor sites. At this point, through electrical activity, the information is transferred."

"So what was Sinclair's teacher telling him?" James asked.

"Well, when I said there are upwards of 10,000 synapses in each neuron, what I didn't mention was that only a little more than 2,000 of them were created at birth. But as the brain experiences the world and learns, and thus begins to grow, the neurons begin sending out and receiving a flood of signals via the dendrites and axons. Therefore, over the first couple years of life, the number of synapses grows tenfold. And they continue to grow at a steady rate through puberty. In fact, every time something is learned during this period, a new synapse is formed. But in early adolescence, the brain begins a pruning process, which I think is what Sinclair's teacher may have been referring to."

"So the brain begins to ... what? ... destroy these synapses?" Kevin asked in an uncertain voice.

"Only those that are weak … and by 'weak' I mean those connections that are not used as often as others. The brain will actually trim away these weaker connections."

"How about neurons themselves?" Kevin wondered.

"Through a different process called apoptosis, even neurons can be short-lived if they are not used frequently. The brain will eliminate them."

"So this synaptic pruning generally takes place in early adolescence?" Kevin repeated.

"Indeed it does."

Nancy leaned forward, studied Rudy for a moment. "So if our synapses are created primarily during childhood and then pruned in puberty, what does that leave for adulthood?"

"That's the hot topic right now," Rudy answered. "Actually it's been a hot topic in some quarters of the neuroscience community for the last several decades," he added. "It was once standard belief that new neurons definitely could not be formed in adulthood and new synaptic connections formed much slower. We needed basically to use and strengthen what we had … become better at what we already could do."

"Well it's definitely more difficult to learn new things in some instances as adults, like learning to speak a new language," James pointed out. "It's much easier at an earlier age. This would seem to indicate that creating a new synaptic connection in this particular area is more difficult for an adult."

"Indeed it does," Rudy nodded, "but the across-the-board belief that new neurons cannot be created is now being seriously questioned. And if new neurons can be created, then so too can new learning—even the total learning of new skills …"

"Neurogenesis," Kevin shook his head in amazement, "even in adults."

"Even in adults," Rudy nodded in confirmation. "And early evidence of this came in the early 1980's by way of, believe it or not, songbirds. It was discovered that songbirds, such as finches, create new songs during each spring mating season. They don't carry over the same songs from the previous year. Basically what researchers found after all the testing on the birds was completed is that the brain cells that create one year's songs actually die off and new brain cells are formed to create the next year's songs."

"So new neurons were being born in these birds?" Nancy asked.

"Correct. And there were other studies that found neurogensis in a variety of rodents as well."

"Was there any experimental evidence for neurogenesis in humans directly?" Kevin asked.

Rudy nodded. "Scientists used something called BrdU, which is a thymidine analog ..."

"Essentially it's a water-soluble compound," James added, "that is incorporated into body tissue to show if new cells have been formed."

"It was used in cancer patients to determine if any new cancer cells had formed," Rudy continued. "Apparently it marks any new cells that are present. Well, researchers in the early 1990's decided to try to use BrdU to determine if new brain cells were present in the human brain."

"There were cancer patients who had BrdU incorporated into their brains," James explained, "to see if the cancer had spread into that region of their bodies. When some of these patients died, neuroscientists dissected their brains, but this time they were not looking for the formation of cancer cells, but for new brain cells."

"And new neurons were indeed located in the hippocampus." Rudy said.

"The BrdU clearly marked them," James added.

"So the old idea of humans being born with a fixed amount of neurons has, over the last few decades, been slowly fading out," Nancy said thoughtfully. "Very interesting."

"Indeed," Rudy said. "And neuroplasticity in general has been tossed around in the neuroscience community for a good period of time. Along with evidence provided by such things as the brain malleability demonstrated by our monks, it was otherwise through animal experiments—in some cases very cruel experiments—that a great deal of evidence was uncovered for neuroplasticity itself. I read recently about some particularly brutal experiments conducted on animals back in the 1980's that really indicated that the brain was indeed capable of rewiring itself."

"The Silver Spring Monkeys from Montgomery County—back in our home state," Rudy said with a shake of the head. "These were experiments that bring up a whole host of ethical questions that we'll have to definitely touch upon another time. But what they showed us seems to confirm that restructuring of the brain can indeed occur."

James leaned forward. "Essentially, scientists surgically severed the region of these monkeys' somatic sensory cortex that was responsible for sending signals from the hand to the brain."

"Basically," Rudy added, "the monkeys felt nothing when their hands were contacted. What they found was that the part of the brain set aside for receiving signals from the hand began receiving them from the feet instead."

"A definite sign of remapping within the brain," James confirmed.

"These tests were taken a step further," Rudy continued, "when one of the monkeys, just before its death many years later, was subjected to a kind of exploratory brain surgery so the scientists could see what was happening in the animal's brain before and after a final set of experiments were performed on it."

"Essentially they were able to examine what exactly went on in the monkey's brain in relation to those severed nerves in its arms. They performed a series of experiments where they would touch different parts of the animal's body to see what area of the brain responded."

"What they expected to see was that there would be absolutely no activity in the portion of the brain responsible for receiving signals from the hand. Since the nerve had been cut, this would be, they believed, a completely silent region. But this wasn't the case, because when the monkey's face was touched, this region of the brain, the region responsible for receiving signals from the *hand*, showed a high level of activity."

"You mean the portion of the brain that was the receiver of sensations from one part of the body was now receiving sensations from another part?" Kevin asked in astonishment.

Rudy nodded. "The scientists believe that so many years had passed since that nerve had been cut that the brain virtually remapped itself. Something that simply was not supposed to take place. Traditional thinking, even at the time, asserted that the brain was suppose to be hardwired … that each section was responsible for receiving signals from specific parts of the body. The brain was not supposed to be malleable in the way that the experiment suggested."

"A complete restructuring had occurred," James continued, "even to the point where the region of the somatic sensory cortex that was responsible for the face region, *and only the face region*, had also begun picking up signals from the hands and arms."

"There were even experiments, again ethically questionable, on animals such as ferrets," Rudy told them. "Since the optical and auditory nerves are similar in ferrets and humans, scientists decided to conduct experiments in these two areas on ferrets. They basically cut the line of communication from one ear in the ferrets to the animal's thalmus. In short, after conducting a series of experiments where light was flashed into the ferrets eyes, they found that the animal's brain compensated by growing branches of the optic and auditory nerves toward one another so that they literally began to share signals. When light was directed into one of the ferret's eyes during one part of the experiment, the ferret's brain had remapped itself to such a degree that the animal actually *heard* the light. It responded as it normally did when it heard something."

"That's pretty wild," Sinclair nodded approvingly.

"You're following all of this, huh?" Nancy smiled at the boy.

Sinclair pushed to his feet and began to remove his camera from the tripod. "What's this whole thing with the brain changing itself called again?"

"Neuroplasticity," his father answered.

Following Sinclair's lead, the adults began to pack up. "This ability of the adult brain to reorganize itself," Nancy said as she eased her backpack over her shoulder, "has so many possibilities ... I mean, everything from education to mental health ..."

"And the fact that observation—whether it be meditation, prayer, trance dancing, or simply intense contemplation—accelerates this remapping, reorganization, and even new growth in the brain, opens up a host of, how can we say, more evolutionary implications," Kevin offered.

Rudy shrugged as he zipped his bag up. "It certainly makes you think. If more and more of the human population actively engaged in a deeper level of observation, both of themselves and of the world around them, what kind of effect would that have on the species as a whole over time?"

"Something to ponder as we continue on our way," Kevin offered with a smile.

◆   ◆   ◆

The trails were now becoming a little bit busier with other hikers. As they left their place beneath the falls, two young couples arrived from one direction, while a group of three arrived from the other.

"Gettin' a little busier, huh?" James said.

"Well, as I mentioned in our email," Kevin said, "they have some activities going on around the park today. In fact, we'll head back toward the campgrounds in a little while to check out the music."

"Yeah, you guys will love the African drum group," Nancy told them. "We saw them about a year ago and they were terrific."

"I want to show you another overlook or two before we head back," Kevin said.

From the falls, the trail angled downward steeply for a good stretch before they came to a break in the tree line. The group paused and Rudy felt a shiver move along his back as he looked down over the expanse of mountain tops that stretched out before them. The sky was so blue that the tree line seemed to take on an aqua texture. One could believe they were standing atop the world.

"Something else, isn't it?" Nancy asked no one in particular.

"Sure is ..." James said, watching as Sinclair quickly focused his camera, quickly snapping at least a dozen pictures.

A little while later they found themselves pushing up "Middle Mountain", again coming upon magnificent views as they moved along. James listened as the other four talked about various things—Nancy's job, Sinclair's photography, Kevin's graphic design business.

As he followed behind the group, James thought about his father and everything they had been talking about throughout the morning. He realized that everything they had been discussing, from conclusions drawn to questions asked could all be, for a lack of a better word, misguided. He smiled, knowing his father would have scolded him for thinking in such a way. "What's important," he remembered his father saying many years before, "is that you are asking the questions, engaging the mind, and examining Nature." His father believed that the intense study of Nature was the only chance humankind had for possibly uncovering the greater mysteries of life and existence. He had told James once that it was akin to coming upon a great river and needing a place to cross to keep your journey moving forward. You may have to search the banks of the river for a very long time before you come to an adequate crossing point. If you choose not to study the river and blindly wade out into it, you may find yourself caught in a current that you cannot free yourself from. Once this happens, there's no turning back and no moving forward. The Rubicon, for all intents and purposes, has been crossed, but you never get to the other side. He feared that humankind was spiritually dangerously close to being trapped within just such a current.

"Come on, Uncle James, pick it up!" Sinclair yelled back to him.

James smiled. "Gotta hate a show-off."

7

April (Afternoon)

The main campground was bustling with activity. There were several vendors selling a variety of foods. In different areas there were musicians, all acoustic, set up and playing music for the hikers. Everyone looked relaxed and content.

Nancy and Kevin had packed a cooler full of sandwiches and sodas. The group gathered at one of the tables, talked, ate, and generally caught up on each other's lives. At one point, a group of about a half-dozen musicians began setting up an array of African drums. One of James' close friends from college had been a musician and had played the kind of drums he was seeing lined up. If he wasn't mistaken, he saw both Djembe drums and Sabar drums being set up.

"This should be good," he said to the group.

"Yeah," Nancy nodded, "it's mesmerizing."

After they ate, the group milled around, chatting with some of the campers. Nancy ran into some people she knew from an old neighborhood where they used to live, and she and Kevin spent a few minutes visiting with them. James walked over to where Rudy and Sinclair were standing. "Rudy, I have something you might want to read."

"Some more of Dad's writings?"

"Yeah, and a part of it has to do with where the conversation left off earlier … when you inquired as to what effect deeper observation might have on the human population over time." James pulled the stapled packet of typed paper from his backpack and handed it to him. "I emailed this to you late last night, but figuring you probably hadn't had a chance to read it yet, I decided to bring it with me." James reached over and turned the pages until he came to the page he was looking for. "Start right here. It's a conversation between two men, one older, one younger. The one speaking in first person is the younger man. They've been discussing the double-slit experiment, so you can just pick up from that point."

"Sounds good," Rudy nodded.

"Why didn't you just pull it up on your iPhone, Uncle James?" Sinclair asked.

"It's a bit large ... and it would be easier to read this way."

"One thing I've been wondering, Dad," Sinclair said, "why didn't grandpa show you many of these writings before he died?"

"I don't know, son" Rudy said quietly. "I don't know."

◆   ◆   ◆

Rudy found a spot under a large tree, the hypnotic rhythm of the drums now pulsating behind him. He turned the packet to where James had showed him, and began reading ...

Sum-over-Paths

Sitting at the small table in front of the fireplace, I look around the tiny cabin. Two entire walls are adorned with pictures of his two daughters and son, as well as his six grandchildren, all of whom now live out of state. But it is the opposite wall that has caught my attention, for it obviously has been reserved for his wife of 48 years who has just recently passed away. He has pictures of her ranging in age from about seventeen to just before her death last summer. The progression of her life had seemingly moved her along in such a gentle way—the essence of who she was had never left her hazel eyes.

I look across the table at the man and see the age that has come to visit him—a wrinkled brow, eyelids sagging precariously over gentle, but intense brown eyes, and the once black beard now gray and fraying at the ends. But his spirit is lively, and his mind sharp. My father told me he had once been a giant of a man, and I believe him.

This is my second visit in less than a week to the cabin, which sits in the woods about a half mile off of the main road. I take a drink from my warm mug of tea and think with a shiver of the long walk in the snow from the end of his driveway.

"Many interesting phenomena have occurred where the electron is concerned," he is saying, "one such phenomenon is what is called the Feynman Perspective, named after the physicist Richard Feynman."

"The Feynman Perspective," I think aloud, "is it also called *sum-over-paths?*"

"It is," he nods. "Basically the double-slit experiment is used, but with a phosphorous screen behind the barrier instead of a photographic plate. What seems to immediately become obvious is that the electrons, as they pass through either one of the open slits are, on some level, aware that the other slit through which they do not pass is open. The subsequent interference pattern that forms is a sign that the electrons, even when they are fired one by one, recognize the circumstance and, in a sense, seek each other out to interact. And the locations where this interaction is occurring determines the locations where the electrons will most *likely* be found. The interaction actually enhances the probability wave."

"That's right. As we talked about before, because of the wave property everything is based on probability. In the same way we can only assign a probability to where a water molecule in a wave will land, we can only assign a probability to where the particle, as part of a wave, will finish its journey."

"Correct," he nods.

"What about where the interaction is *not* occurring?"

"The probability wave is lessened at these positions."

"So the likelihood of finding the electrons there is also lessened?"

"Correct," he answers as he holds up a finger in the dim light, "but here's where it starts to get strange. Feynman approached things differently at this point because he actually spoke out against the notion that the individually fired electron goes through just one or the other of the open slits."

"Don't tell me he proposed that each individual electron somehow goes through *both* slits?"

"That's exactly what he proposed."

"But that's crazy."

"Our common sense and experience indeed tells us that they must go through one slit or the other, but as we've discussed, by observing we have altered the experiment. 'Cause remember, to observe the electron we have to engage it and that means, in most instances, using light as the detection device. But by using light as the mode of observation means bringing photons into the mix. Photons bounce off of material objects all the time in our macro world and cause no effect on their motion. Photons hit the car you drive, but the motion of the car is not affected. They hit and bounce off of the wall, but cause no effect on the motion of the wall. But the motions of electrons are affected by photons."

I nod, "And altering their motion alters the experiment."

"Exactly."

"How?"

"Well," he shifts in his seat, "once the quantum domain senses that it has been determined that the electron definitely passed through one slit or the other, it somehow signals to the electrons to act differently. At this point, the electrons will continue to travel through the slits but now will not interfere with one another. They will, instead, hit the screen *directly* behind the slit through which they pass, like one straight beam."

"They suddenly begin to act as fired *particles* again, and not as waves. As we discussed earlier, almost as if they shift gears according to the new circumstances."

"And the *circumstance*, in this case, is that they sense the observation and this alters them, thus making verification that they actually travel through one slit or the other virtually impossible."

"That's mind boggling," I shake my head. "So Feynman proposed what?"

"That this lack of our being able to verify that they travel through one slit or the other, seriously opens up the possibility that each electron travels through *both* slits."

I rub my eyes with both palms, shake my head. "It goes against classical common sense."

He holds up a finger. "But things get stranger still. Feynman went on to propose that when going from point A, which is being fired from the point of origin, to point B, which is the final destination on the screen, each individual electron actually travels *every possible path at the same time* to get to where it is going."

"Come again?"

"It travels at the same time each and every possible path in its journey to strike that exact same point on the screen. It searches out and takes every possible path to get it from its starting point to its point of destination. And it's an infinite number of paths. Think of crossing a five acre open field to get to a specific place on the other side. Imagine how many different paths you could potentially take to get there. I mean, it's unending."

I almost laugh in my amazement and uncertainty. "How is something like this possible?"

He stares at me, presses his hands together. "Remember, we are learning all the time that things, at a level beyond our ability to see them, act completely different from what we are fully able and used to comprehending."

I sigh. "I know, but it's so … unfamiliar and strange."

"That, my young friend, is the quantum world."

"But I still must ask the question: how?"

"Well," he sits back, "Feynman was able to work out the math to verify his prediction. Without going into all the specifics, he essentially concluded in his mathematical formulation that all the paths, after being combined, ultimately wind up canceling each other out so that only one of the infinite paths winds up being significant when it comes to the motion of the object. As we mentioned before, this approach is called *sum-over-paths*."

"So, let me get this right in regard to the double-slit experiment. If we were to witness the experiment as you describe with the naked eye, we would only see the end result, which is the beam hitting that one spot on the screen. We would, obviously, neither be seeing all of what is happening at the most basic level, nor *how* it is happening. And the what and the how defies what we hold as common sense."

"And it defies classical physics."

I sit back and sigh. "I can't help but now wonder just how many other things that we view or experience are based upon something else, or some other process entirely different from what we think or reason?"

"Quite a bit, I would imagine, quite a bit." Then he leans toward me. "But our everyday lives and world are seemingly not affected. The tennis ball hit over the net, though going through all of these things at the subatomic level, still follows its same trajectory as we have come to know it."

"But why, I wonder, do these electrons feel compelled to move along every possible path like that?" I shake my head. "That question alone amazes me, not to mention again bringing me back to the question of consciousness."

"Speaking of consciousness, let us return for a moment to the tennis ball I just brought up. As noted, the trajectory doesn't change, but why doesn't the trajectory change if all potential paths are not only within the subatomic mix, but are in fact traversed?"

"Because of the object's size?"

"In part yes," he acknowledges, "but even with the small particles, like in the aforementioned double-slit experiment, the final destination is not altered as well."

I stare at him. "It's almost as if, on the subatomic level, things work together to present us the world in a way we are able to comprehend it." I pause, almost as if not believing my next thought, "if we evolve our view, our perspective, our understanding of life, the world, existence—would it then maybe alter how it presents itself to us?"

Event Horizon

Sliding my chair away from the table, I place my elbows upon my knees, cup the top of my head in both hands. "Okay, let's pause here for a minute," I request. "I need to try and make some sense out of all of this."

"Practical application?" he asks.

I shrug. "As far as technology is concerned, the potential for all sorts of practical applications is apparent, but what does it all mean on another level is what I guess I'm struggling over."

"What is this *other* level?"

I push to my feet, returning again to the window. "I guess in some way," I admit, staring out at the quickening pace of the snowfall, "that's probably the most logical first question. What exactly am I talking about?"

"Spiritual?"

I sigh, watching my breath cloud once again upon the cold glass. "I'm never satisfied with the language we use for such things, or for the perception we have built over time for the meanings for certain terms … but for expository purposes, I guess we can, for now, speak about a 'spiritual' level."

"From your perspective, what exactly is this level, this realm?"

"The first thing that comes to mind is what we call the 'soul'."

"What is the soul to you?"

"To me, the soul represents your deepest level of consciousness."

"Here we are back to your consciousness question."

I turn to look at him. "And it's a question, sadly to say, that I'm not yet equipped to answer. Not even close."

"That's okay," he assures me, a gentle, nonjudgmental smile upon his lips. "It's something that you may need to work up to."

"My event horizon," I quip.

He nods. "Possibly. But like I say, let's stay where we are for a few moments. Remember every branch leads somewhere," he reminds me. "Let's continue with the soul. How is the soul level, the deepest level of consciousness, as you call it, reached?"

I turn back to the window, noticing the perfect shape of each and every flake as they land upon the windowpane; for some reason, it makes me think of how beautiful mathematics must be to someone who truly understands it. "I need to think long and hard about that question," I say as I turn back to study him again. "How would you say it is reached?" I ask. "Where would the beginning lie?"

He thinks for a moment. "First and foremost—curiosity. Curiosity may be one of the universal passageways."

"Hmm," I nod, lean my back against the wall, feeling the cold draft from the nearby window. "Then what's the initial manifestation of curiosity?"

"Observation," he answers without hesitation.

Letting out a deep breath, I press my head against the wall and notice how the candle's glow breaks the room's darkness perfectly in two. "You know, I have this crazy theory of sorts circling through my head," I admit. "But it's still not even close to being fully formulated and it may be way out there."

"What is it?"

I hesitate, shake my head. "It's not ..." I begin, but pause.

He lets out a sigh of his own. "Don't begin to justify it," he tells me, "just place your thoughts into the open. Whatever you're thinking will surely lead you somewhere, even if it's not exactly where you think it will."

I turn back to the window. "With all of this talk of observation, of measurement, particularly in regard to particles ... and even now with our brief discussion of the soul ... it just has me thinking, that's all."

"About consciousness?"

I nod again. "It's constantly stirring around in my head, but it's at this point still just a skeleton."

"*Let it out*," his voice is patient, but firm, "stop worrying about if it's right or wrong, complete or not and just put it out there. Maybe we'll be able to fill in the gaps later, who knows? But for now just say what you're thinking—pride and reluctance have no place in this pursuit."

"I have," I let it out, "a feeling, an intuitive sense I guess it is, that consciousness, in the end, might just wind up being the foundation of all being, even the world of matter." I pause, begin to pace back and forth in front of the window.

"Go on," he urges.

"With what has been determined about particles and observation," I continue, "with everything from some particles having only a potential existence until observed, to light varying between wave and particle depending upon the circumstance and the kind of observation taking place, to the apparent communication that occurs between quantum entangled particles when one or the other of the pair is measured or observed, to Heisenberg's uncertainty principle ..." I stop my pacing, stare down at him in the dim light, "and I could go on and on with examples of, what appears to be, the involvement of consciousness ..."

He nods. "I'm following you. Keep going."

"And how we determined," I stay in stride, "particularly through the double-slit experiment, how the particles, in a sense, seem to conspire to present the world to us in a way we can understand, regardless of what is really happening at the basic level. The particles travel through both slits and create an interference pattern on the screen, but yet we only see what appears to be two beams, one through one slit, one through the other, both subsequently striking the screen in a way that makes sense to us." I sit down in my seat across from him. "If you will, allow me a few minutes of rambling speculation ..."

"That's what I'm talking about. Go ahead."

"What would happen," I continue on, "if our perception of the world, particularly through the study of the quantum domain, truly changed because of our increased knowledge of it? What would happen if we no longer expected to see, for example, two beams of light in the double-slit experiment, but we truly expected to see the interference pattern instead? Would the quantum world ever possibly react to this new perception, this new dimension of our observation, and eventually alter the way it presents itself to us?"

"Allow me now to speculate as well for a few minutes in the direction you have pointed us," he requests.

I nod.

"I imagine," he begins, "that we would, on our end, have to do more than just enhance our knowledge. That seems to me to be only half of the equation. Something more would have to take place within for us to evolve to that next level."

"What?"

"Well, we spoke about curiosity manifesting itself through observation, but what would be the basic component of this 'observation' we find ourselves so often talking about? What would be the first step, *within us*, that would cause our natural abilities for observation to evolve?"

I think for a moment. "We put curiosity in motion through observation," I hear myself thinking aloud, "but what would put observation in motion?" I pause, stare straight into his eyes. "Awareness," I utter the word before practically even thinking it. "*A heightened awareness.*"

He nods. "To observe at the level of which you are now speaking, I would imagine we would indeed need to evolve in this area as well."

"I would ask how," I say, "but I think I know? Through observation, via meditation, prayer, and the other methods of contemplation we spoke about earlier ..."

He nods. "After all, observation at this level serves, above all else, to quiet the mind, thereby freeing it from the clutter of incessant self-chatter." He shrugs, "Not to mention what it apparently does to the structure of the brain."

"Pure awareness."

"Correct. And once the mind is still and reaches the Clear, one finds that they are perfectly in the moment, directly experiencing life and existence as one thing unfolding within them and about them. At this point, one is perfectly open and receptive, without judgment, without opinion, without conditioning, without constraints."

"Experiencing life and existence as one thing unfolding within them and about them," I repeat his words. "A more open, receptive, clear *observer*?"

"I would say so. And couple it with a greater overall knowledge and understanding, along with the impact this also has on the brain via neurogenesis and the like," he shrugs, "who knows?"

"I don't know, but my next thought is not even close to being formulated, but here goes: to cause evolution in the reaction by the physical world to this new level of observation, would this new level of observation and understanding need to occur in great numbers to trigger it?"

"A critical mass, of sorts?"

"I guess that's what I'm asking."

"Hmm," he sits back, stares off in the middle distance. "I don't know," he says after a few moments, "I really don't know." Then he leans forward again. "I imagine it might if we are to have it become a part of our tangible, everyday world ... the world that is experienced and taken as our reality. Evolutionary changes in a species generally take place when the whole species experiences the need for that change over a sustained period. For example, we developed at some point the need to stand upright, but surely this change only began to occur after the whole, or close to the whole, species found itself with such a need. Evolution does not happen in a vacuum, in isolation. Reaching bipedalism was certainly a long process for humankind; Nature did not respond instantaneously, I'm sure, after the first few individuals began to experience the need for upright mobility. In matters of evolution, Nature responds to the sum of a species; evolutionary changes of the mind, heart and understanding, therefore, must occur in the same manner."

"So one cannot get to this next level on their own, as an individual entity?"

"Maybe they can," he answers, "but I would think only to a certain degree. I imagine it would be fleeting and not part of their everyday world. They would still have to deal with day in and day out reality on the physical realm, which Nature provides for the whole of the species. But for Nature to respond to the species in its totality," he reiterates, "I would think the species as a whole or near to a whole, would need to evolve in both their level of understanding and their level of awareness. So it, the knowledge and awareness, would become almost like that of an instinct, an intrinsic part of our genetic make-up. After all, if all is indeed one, as the unified field suggests, then to move forward as that one entity, we would need a very large majority of our species to evolve to that level."

"Something you just said is very fascinating," I tell him as I sit back in my chair and cross my arms over my chest. "Actually several things you said caught my attention. First, when you said that it needs to become like that 'of an instinct, an intrinsic part of our genetic make-up', three letters popped into my mind."

He holds up his hand. "Let me guess," he says, "DNA?"

"Genetics," I continue in stride, "most certainly holds many keys. In fact, though I know virtually nothing about it, I've often thought that maybe DNA already has these things, or the potential for these things, wired within us, even our discovery of DNA itself. If everything is part of the unified field, then do we not, at the molecular level, share the potential for virtually anything that exists? Is everything hard-wired within us at some level, and is it a matter of us evolving to them?"

He sits back, lays his hands in his lap, and stares intently at the tabletop. "Interesting."

"But let's leave genetics for another time," I say as I begin to pace. "The second thing you said that caught my attention was the fact that an individual has to live with the reality of his species. It just made me think that to evolve spiritually is the most difficult *physical* thing you could ever do. Your physical being is your greatest limitation, from your desires and hungers to the limited scope of your five senses. And that leads me to another sudden, not thought-out question. Speaking of evolution, does Nature change how it presents itself to us, or does it alter us so we can understand it as it actually is? Do we adapt to it, with its help, or does it adapt to us?"

"Are you speaking about change to our understanding, our awareness?" he asks. "Or of evolution in general?"

"Our understanding and awareness."

He nods, thinks for a moment before answering. "Probably all of the above," he begins slowly, "but on the surface, at least in regard to what we can see, it would probably need us to change initially ... to this end, when we are ready, maybe it would evolve in us a heightened sense, or maybe even an extra sense, so our position, our perspective would change. From there, it wouldn't surprise me if Nature began to adjust and readjust as we evolved." He smiles with a shake of the head. "The continually opening and closing Pandora's box."

"In essence then," I turn to face him, "it would continually be forcing us to evolve."

"And we it," he adds.

"With the process beginning again each time."

"It would have never stopped to begin with."

I return to my pacing. "How would that next level, that new perception be?"

"Beautiful," he says softly, "most certainly beautiful."

I sit down again in front of him, but this time perched on the edge of my chair. "So for one to begin to experience the world on this new level, the species *as a whole, or nearly a whole,* would need to evolve first?"

"I would think."

I sit back, almost in defeat. "But how?"

He lets out a sigh, shakes his head.

I stare at him. "I can't help but now think of Plato and his shadow world. Could it be that the world of shadows he referred to may currently be our greatest obstacle in regard to our species' 'consciousness' evolution?"

He shrugs. "Could be. It very well could be."

I stare at him. "Let me put you on the spot."

"Sure."

"If you had to suggest one thing, to implement one step in a plan to begin modern-day man on his way to this kind of evolution, what would it be?"

He laughs. "That's some question, but I'll try to give you an answer." He sits forward, rests his elbows on his knees and cups his chin in his palms. "Because of the hectic nature of modern life, it would have to be something subtle, something on a small scale." He stares past me for a long time, then sits back, stares out toward the window. "I've always wondered what kind of effect it would have if everyone who was able, and by 'able' I mean those of us who do not have to worry about meeting basic survival needs. If everyone who was able," he repeats, "took it upon themselves to set aside just twenty minutes each day. A half hour would be more ideal, but just twenty minutes would suffice. The first ten minutes of this twenty would be engaged in the learning of something they did not know about the natural world. It could be about plants, animals, space, human cells … it could be about anything at all that resided in the natural domain. But every day, it would have to be something new. The second ten minutes, immediately following, would be spent in some kind of meditation. It would be up to the individual to decide which form of meditation." He sits back with a smile. "There you go."

I stare at him. "That would be your first step?"

"That's it. Not exactly Einstein's theory of relativity, I admit."

Tired now, I drop my head back and stare up at the ceiling. "It's definitely practical and doable by everyone."

"Think about it for a minute though," he encourages. "I always thought it would be an interesting experiment of sorts, especially after a year or two of this had gone by. In how many people would the wonder and mystery of life, even in a tiny way, be awakened? Would any level of curiosity, even if just subtly, have been sparked?"

I ease to my feet, amble back over to the window. The night sky has cleared into a kind of purple expanse, and just a sliver of moon hangs low in the sky. "What's next for us?" I ask him.

"Only tomorrow will tell," he says, "only tomorrow will tell."

8

Rudy pauses, lies back onto the cool earth and listens to the pounding beat of the drums. His mind works over everything he has just read. After a few minutes pass, he sits up and continues ...

The Einstein-Rosen Bridge

Reflecting away watery sunlight, a blanket of snow lies untouched before me. I lean into the cold winter breeze, push on toward the lone cabin that sits at the very end of this now hidden driveway. Shoving gloved hands further into the pockets of my leather coat, I bend forward, lengthen my strides and think of just how beautiful the mystery of it all actually is.

As I approach the small dwelling, a path of broken snow trailing behind me, I sense a presence at the edge of the woods just off to my right. Pausing, I catch a glimpse of a tiny brown rabbit perched upon back legs, peeking at me over a small bush. If it wasn't for the slight twitch of his ears, I would have never even noticed him. I stand perfectly still and watch him. He, in turn, stands frozen, watching me. After a minute or so, I smile and start again on my way; out of the corner of my eye, I see him ease down and hop away into the forest. I shiver and begin to quicken my stride, but am quickly startled by what sounds like a loud hiss coming from the forest. I turn just in time to see the slim figure of a fox take off from out of the forest's underbrush, the rabbit hopping frantically just a few strides before him. With amazing speed and agility, their chase makes a giant figure eight in the unbroken snow before they, in a matter of seconds, disappear back into the forest. My heart pounding now, I watch and listen intently, but hear and see nothing. I wait several minutes, but still there is nothing, just the winter silence and stillness resumed. I pull up my collar and push on, not quite knowing what to make of what I just witnessed.

◆   ◆   ◆

"I wonder if my little rabbit friend escaped?" I ask after settling down at my place at the table, a steaming mug of hot chocolate sitting before me.

He sets a dry log onto the top of an already crackling fire, pushes to his feet, walks back across the room to his seat across from me. He pours himself a cup of hot tea, sits back and takes a long sip.

"You couldn't see the conclusion of the chase, huh?"

"In a flash, they were gone … back into the woods. It was quite amazing to watch."

"From what you saw, what do you think happened?"

I shake my head. "It happened so quickly … it's hard to tell. It's fifty-fifty."

"Fifty percent chance that he's still alive?"

I nod. "And, sadly, fifty percent chance of the alternative."

He sets the cup back onto the table. "For some reason, your rabbit story makes me think of *Schrodinger's cat*. The whole collapsing of the wave function thing …"

"How does this apply to my rabbit?"

He sits forward. "Granted, this scenario is a bit different in that we have the observation of the fox to contend with. But let's say, just for example, that the fox caught your rabbit, but the rabbit, though now badly injured, was able to get away and make it into the safety of his hole. For the sake of discussion, this scenario eliminates the fox's observation. At this time for the rabbit there's basically a fifty-fifty chance of him surviving his wounds through the night. At the end of the night, would he be dead or alive?"

"From the quantum perspective," I answer, "he would be both dead and alive until someone dug up the hole and made a determination." I shake my head. "Man, that's bizarre."

"It is bizarre," he agrees. "But let's look at some examples taken directly from the quantum world itself. Think back to the double-slit experiment and how the photons alternated back and forth between wave and particle, depending upon the kind of observation/measurement being performed. And think of how a single photon can be in two places simultaneously, as was illustrated when both slits were open. In the same way that your rabbit is said to be both dead and alive, the photon, as its behavior in the experiment shows us, illustrates the reality of all of this on the quantum level."

"But how?"

He sits back. "That's one of the questions. Along with *where?*"

"What do you mean *where?*"

"Well, we're talking about the wave-particle duality that exists, and not only with light, but with all matter, even your rabbit. This explains in part how the particles behave when confronted with the choice offered by the various slit options. As we just stated, they behave as a wave in some instances and as particles in others. But in a way, this only tells part of the story, particularly if you subscribe to the notion that one particle goes through *both* slits simultaneously."

"Similar to my rabbit being both alive and dead simultaneously?"

He nods. "If you don't agree entirely with the theory that the individual photons are simply recognizing the situation and reacting to it in a discerning way to create the various patterns on the screen, as we discussed at length earlier, then the other explanation you find yourself facing is the possibility of branching universes. What is also referred to as the multiverse."

"So what you're talking about are parallel universes?"

He nods again. "But let's first start with an interesting set of questions that have cropped up thus far. First, if two choices, two options cause particles to affect each other, to *interfere* with each other, as is demonstrated in the double-slit experiment, then do these two options both exist simultaneously? Secondly, if the sum-over-paths phenomenon is indeed true, if the possibilities can somehow add up as they appear to, then do these possibilities *actually* exist somewhere else as well?"

"So I guess the next question you're leading up to is: does the particle, and my dead/alive rabbit for that matter, exist in both places, in both 'options', at the same time too?"

"That may indeed be the next in a long line of questions. And if they do exist simultaneously, then that leads us to our question of *where?* Where are they existing?"

"Hmm," I sit back, let out a sigh. "And this is where parallel universes come into play."

He sets his mug down. "This is one of the theories bouncing around. There are many names for it out there right now, none of which I really care for. One name, as I mentioned, is the *branching universe theory.*"

"I've also heard the *many worlds interpretation.*" I add.

He nods. "It's sometimes called that too."

"So what physicists are talking about is, in essence, the existence of overlapping parallel universes?"

"Basically, yes. I like to refer to them as relative states instead, which I procured from John Wheeler, the now legendary Princeton physicist. He referred to the parallel universe theory as the *relative state formulation*. I like to call them *r-states* for short."

"So, from what I can glean from all that we've been saying here, the theory must be proposing that, in the case of the double-slit experiment for example, the photon, or the electron, actually branches off into a separate world? A parallel universe?"

He smiles. "A relative state, or r-state."

I nod and correct myself. "A relative state."

"And then the two r-states interact," he continues my thought, "in essence, they overlap and interact to form the one reality we experience. In this case, the interference pattern on the screen, or with your rabbit ... as it takes on the state of being dead or alive."

"So in one r-state the two worlds interact and live out the one possibility of my rabbit being dead, while in the other r-state they live out the possibility of him being alive."

"And then the two merge back together to give us our reality of the animal being either alive or dead."

I sit back, shake my head in disbelief. "That's so bizarre, so far out there." Then I think it through for a moment. "But it does give a framework for how the phenomenon can take place—how the particle can be in two places at one time." Then I look at him closely. "And physicists are seriously looking at this theory?"

"It's one of the theories currently on the table. In fact, the theory was first legitimately put forth back in the late 1950's by a young physicist from Princeton named Hugh Everett III. And modern, highly regarded physicists such as David Deutsch and John Wheeler, among others, have not ruled it out. In fact, David Deutsch, of Oxford University, is currently one of its biggest proponents—he refers to it as the *multiverse*. For the sake of consistency, I like to call it the *multi-state*."

I hold up a hand, letting out a slight laugh. "Okay, let's pause here for a few moments and break this down."

"Sure."

"Where can someone begin if they really want to understand this whole thing?" I wonder.

He takes a long sip from the mug, obviously gathering his thoughts. "I guess a good place to begin this particular discussion would be with the very large and then work our way down."

"Are you speaking of the universe?"

He nods. "That might be a good place to start." He sets the mug down. "Oftentimes the sheer vastness of the universe is hard for people to grasp." He adjusts himself in his seat. "And to get a true perspective, we need to first preface it all by having some sort of understanding regarding just how fast the speed of light actually is." He snaps his fingers. "Just that fast," he says, motioning toward the sound. "In the time it takes to snap your fingers, one traveling the speed of light would have circled the Earth approximately seven times."

"Isn't the speed of light something like 186,000 miles per second?"

"It is."

"So if something is a light-year away," I clarify, "then you would need to travel, *at the speed you just indicated—the snap of your fingers*, for an entire year to reach it?"

"Correct. And with this said," he continues, "one of the best descriptions, visually speaking, that I have ever encountered in regard to explaining the size of the universe was put forth by the author William Barrett in his book, *Death of the Soul*. Again, having read it so many times, I know it verbatim." Closing his eyes to better recall the words, he begins to recite:

"… We, on this Earth of ours, are embedded in the great system of stars called our galaxy. Shaped somewhat like a pocket watch, the galaxy is about a hundred thousand light-years in diameter, perhaps fifteen thousand light-years thick, and contains approximately a hundred billion stars.

… Copernicus dislodged the Earth from the center and showed it revolving about the sun. But this sun of ours, a small star, itself revolves with other stars within its galaxy, completing a revolution about once every two hundred fifty million years. But the vistas of distance only begin here. We can think of our galaxy (with its hundred billion stars) as our 'island' in the ocean of emptiness which is the whole universe. There are millions of other 'islands' known, and undoubtedly countless millions of others not yet seen. Each galaxy consists of billions of stars and each is some one or two million light-years from its nearest neighbors.

And so on, and so forth, vista upon vista and motion upon interlocking motion—until the mind reels before this picture it cannot assemble."

I nod slowly, understanding why he thought the description was so compelling. "So our galaxy is just one … of millions of galaxies?"

"That's correct."

"And each galaxy is numerous light-years from the other?"

"And just think," he snaps his fingers again. "You would need to travel this fast for a hundred thousand years just to cross our galaxy."

"And then," I add, "one or two million years to reach the next galaxy ... again," I nod toward his fingers, "traveling that fast."

"And this would not even begin to scratch the surface for, again, there are *millions* of galaxies, some far bigger than our own. And on top of it all, our universe, from all indications, is expanding and infinite."

I sit back. "Amazing."

"Now," he continues right in stride, "as we talk about speed, velocity, and indeed time, we need to look for a moment at Einstein's theory of special relativity. If you're moving alongside an object traveling at any speed, the object looks as if it is standing still. You really can't tell if you're moving or it's moving. The truth is that you're both moving relative to the other. Both frames of reference are equally correct—this is the underlying point of the theory of special relativity. One frame of reference is not 'right' and the other 'wrong'. We need to state this up front before continuing. Now, with this said, from *your frame of reference*, relative to you, the object is 'moving'. And time is affected accordingly. An example would be measuring the time it takes a fast moving vehicle to get from one point to another, let's say a distance of fifty feet. If you had two people measuring with two precise stopwatches, one person standing on the ground beside the vehicle, and one inside the vehicle, you would find a discrepancy, though very slight, in the times measured. It would be minute, but the time shown on the watch measuring from inside the car would be slightly behind the stationary watch measuring from outside the car."

"Isn't that called time dilation?"

"Yes."

"So in other words, the faster you are moving relative to something else, in this case the person inside the vehicle relative to the person outside the vehicle, the slower time moves in relation to that something else?"

"In essence—yes. 'Motion' apparently affects the measurement of time. And as you said, the faster you move *relative* to something else, the measurement of time of an event from the two frames of reference will be different. The measurement from within the frame of reference of the one that is 'moving' will record a shorter measurement of time. So, taken to the extreme, if one of these vehicles were to somehow reach the speed of light, the measurement of the time elapsed would be zero ... as if the clock had stopped."

"So speed affects time, and it's all *relative* to the observer, to the position of the measurement."

"And what's more, speed apparently affects the moving object directly."

"How so?"

"Well, if you took the same scenario, and along with the stopwatches used to measure the time, you added a ruler to measure the length of the vehicle. If two measurements were taken, one before the vehicle was moving, and one after it was in motion, you would find a discrepancy in the vehicle's length. Though it would be difficult to take the measurement by this same method while the vehicle was moving, this measurement could, however, be reached in a roundabout way. The stationary observer could take his stopwatch and time the moving vehicle as it passed by, starting the watch as soon as the front of the vehicle reached him and turning it off precisely when the back passed him. By then multiplying the elapsed time by the speed, the length of the vehicle could be determined. Assuming no mistakes were made, and everything was done precisely, you would find that the moving vehicle actually had a shorter length than it had when it was stationary."

"Explain."

"Well, since the speed of light is constant, Einstein concluded that this could only be happening because the speed was causing the vehicle to actually grow shorter, in essence, contracting in the direction of its movement."

"What would happen if the vehicle was, again, somehow to reach the speed of light?"

"According to the stationary observer's measurement, it would grow continually shorter the faster the vehicle moved, until it would disappear altogether when the speed of light was reached."

"So at the speed of light, time would stop and the vehicle would disappear?"

"Correct. But," he holds up a hand, "this would only be apparent to observers viewing it from a *stationary* perspective outside the vehicle. For someone moving right along with the vehicle, nothing would seem unusual. To the individual inside the vehicle, the ruler would appear unaltered, and all would seem normal."

"Special relativity."

He nods. "Indeed."

"Isn't there something in relation to this called the twin paradox?" I ask.

He nods. "Einstein used the twin paradox to basically explain the same thing in a different way. The scenario begins with two twin girls who are 30 years old. One twin is placed in a spaceship that is capable of approaching the speed of light, while the other twin is left back on Earth. Let's say, from the *Earth's per-*

spective, the space ship travels at eighty percent the speed of light and goes out into space for 12.5 years. Once it reaches the 12.5 year point it turns around and comes back. When it lands on Earth, the twin inside the ship is 45 years old, but her earthbound twin is 55. On Earth, 25 years would have passed, but only 15 years within the ship."

I sit back, and take it all in for a few moments. Then I sit forward again. "I know this may be a little off of what we're talking about, but why is it not possible for an object to reach the speed of light?"

"Well basically," he answers, "the mass of a moving object increases with its velocity. See, near the speed of light, an object's mass approaches infinity. Therefore, in essence, it would take an infinite amount of energy to accelerate an infinite mass." He takes a long sip of tea. "But this is where the quantum realm, with its zero mass particles, comes into play, all of which we'll be getting into shortly."

Mug in hand, I slide my chair away from the table, stretch my legs out before me. "So time's relative to the observer?" I speak more to myself than to him. "My consciousness question is trying to sneak back into the equation again."

He smiles. "Time is still a great unknown in many respects. Does it indeed flow forward from a past into a future, which is how we experience it, or does it not flow at all?"

I push out of the chair and, for the first time during this visit, make my way over to the window, the warm mug clutched in both hands. The evening is falling quickly, and as my eyes scan the still untouched snow, I can't help pondering the fate of my rabbit. "Almost every night," I begin to tell him, "right before I fall asleep, you know, right at that moment, that instant, where you begin to drift off, I sometimes have intense revelations that jolt me awake. Many times I find them so intriguing that I crawl out of bed in the dark and write them down."

"That's interesting," he nods. "Did you know that, in some sects of Buddhism, that exact moment you just described, right as you enter sleep, is considered an extremely deep and natural meditative state? Having insights, visions, etc … at that precise moment is not nearly as far-fetched as it may sound."

I look back at him. "I never knew that."

He nods. "For those who experience it, it can bring great moments of revelation."

I turn back toward the window and stare off toward the forest, now fading away into nightfall. "I've had three in the last few weeks. The first of which, strangely enough, is sort of in line with what you just mentioned regarding how time may not flow at all. It was maybe two weeks ago when I awakened to the thought that *time is one thing*. In fact, that's exactly what I scribbled down in the dark: *time is one thing*."

"Hmm," he sits back, nods slowly. "What do you think you meant by that?"

"Though I was sure at that precise moment," I answer pensively, "it left me as quickly as it had come. But nonetheless, I still believe in the validity of the words—particularly because the feeling they left me with when I awoke was so glorious, so clear. It was a feeling of revelation. And even though the clarity of their meaning has left me, this hasn't diminished my eagerness to explore them. The thought they represent is like a beacon of sorts. But," I turn back to him, "I'm not now sure what it completely means ... or how to get back to it."

"Start small and work out?"

I sigh. "Well, on a most basic level, I know that the future and the past are not what I have always believed they are, nor are they what I have experienced them to be. In fact, I'm not sure they even exist."

"Go on."

I lean against the wall, hold the mug up against my forehead, let the warmth relax me. "There may in fact be only the present. What we perceive as being the future and the past is just an extension of the present ... *is* the present."

"The eternal now, of sorts."

I nod. "Of sorts, yes. And it all comes down to where we find ourselves in relation to what we *perceive* as the dimension of time. Our position determines it."

"Not the other way around?"

"You mean, where we find ourselves in relation to time, or to the *flow of time*, determines us?"

He nods.

"That's how I think I've intuitively always thought of time, and our relationship to it, but now I'm not so sure. I mean, it seems to me that the present, the future, and the past all meet in one place. And that place is the exact position of the observer."

"The merging of space and time."

I move across the room, sit down again. "Since that night, Einstein's theory of space-time has intrigued me. I went to the library the next day and read about it for the first time."

He nods in affirmation. "Along with the three spatial dimensions," he tells me, "Einstein added *time* as the fourth dimension. It revolutionized our view of the universe."

"Space and time as one continuum," I utter it to myself, my next thought already formulating. "Didn't he also establish the link between gravity and acceleration ... basically concluding that there really is no difference between the two?"

He nods. "It's appropriate that you mention gravity at this point. Gravity's the great and mysterious wildcard. And yes, in another insight of his, he established the link between gravity and acceleration. But before we delve into the relationship between these two, we probably should look at the evolution of our knowledge concerning gravity." He adjusts in his seat. "To do this right," he says, "we should really start from the beginning." He takes a long sip from his mug before proceeding. "Newton was really the first to delve into the question of gravity. Basically he wondered how two bodies that are physically separate from each other have an influence upon each other. Clearly, he was one of the first to understand that everything exerts a mysterious force on everything else. A force of apparent attraction. And what's more, everything not only exerts a force on everything else, but also *feels* a force from everything else. This force, of course, he called gravity. It was because of this force that your rabbit, just for an example, as well as all other material things, including us, stays fixed to the Earth's surface. This force explains why the moon maintains its orbit about the Earth, and the Earth about the sun. Everyone, of course, has heard the story about Newton's apple falling from the tree. He realized that some force from the Earth pulled the apple to it."

"Didn't he believe that if one of the separate bodies was somehow removed, the other body would feel an instantaneous effect from the resultant change in the gravity?"

"He did indeed. He felt, for instance, that if the Earth somehow vanished, the moon would feel an *instant* effect from the change in their mutual gravitational attraction. It would, among other reactions, most certainly fall out of its orbit."

"And wasn't this the part of Newton's theory, the instantaneous effect, that troubled Einstein some three hundred years later?"

He nods. "It was indeed. This one point contradicted Einstein's theory of special relativity, in that Einstein had already determined in his theory that nothing can travel faster than the speed of light. Therefore, this *immediate action, this immediate signaling* in Newton's theory rubbed against the grain for Einstein."

"It was at this point, if I recall, that Einstein really began to study the question of gravity."

"It was. In fact, his first insight had to do with the link you just mentioned between gravity and acceleration." He adjusts in his seat again. "To explain the link in clearer terms, imagine you find yourself kidnapped by aliens, aliens who are interested in knowing, among other things, the effects of space travel on the weight of earthlings. Just before takeoff, you find yourself tied to a scale. As the ship blasts off, you feel an immediate tug against the scale as the ship accelerates upward to break into outer space. The sensation reminds you of your last roller coaster ride at the amusement park, when you felt a jolt as your back pressed against the back of the seat as the roller coaster accelerated up the track. And now, as you feel the same pressing sensation against the scale, you see your captors noting that your weight has increased. However, a little while later, as the ship moves farther away from the Earth and its gravitational attraction, you see that your weight is decreasing. At this point, the aliens suddenly turn the ship around and a little while later you find yourself landing again on Earth. At this point, instead of freeing you, the aliens want to do one more test. They want to take off again, but this time they want to see if they can somehow accelerate the ship in such a way that they can maintain a steady body weight for their captured human. In essence, they want to see if they can balance out the effects of gravity by carefully altering the moment to moment acceleration of the ship, thereby being able to lessen the fluctuations in your weight. As you take off again, you notice that the ship, still feeling the full force of the Earth's gravity, accelerates, but not as aggressively as it did during the first takeoff. You note, along with your captors, that indeed your body weight did not increase nearly as much as it did during the first flight. There wasn't nearly the same jolting and pressing feeling as there was the first time around. With a sigh, you wait until farther into your trip when the gravitational force from Earth begins to weaken. You remember how your weight decreased during this point during the first go-around. This time, however, just when your weight begins to decrease, your captors increase the ship's acceleration. You begin to feel a slight tug against the scale as the acceleration compensates for the loss in gravity. Because of this ploy engaged by the ship's alien captain, your body presses against the scale just enough to keep your weight constant. Gravity and acceleration, in effect, have canceled each other out. With the expected link between acceleration and gravity once again confirmed for the aliens, you are kindly returned to Earth."

"Quite a story," I grin.

He laughs. "It was the best I could come up with on short notice. But I think it illustrates the point. In the story I just told, it would be difficult, if not impossible, to know if it was gravity or acceleration causing the pulling you felt during your spaceship ride. This interwoven relationship, this *indistinguishability* between gravity and acceleration Einstein called the equivalency principle."

"Isn't this a primary component of general relativity?"

"Most certainly. And discovering the acceleration and gravity link was very important in Einstein's mind. See, since accelerated motion is tangible, he felt that it could really be used as a vehicle to better understand the completely mysterious force of gravity." He pauses, holds up a hand. "But to use acceleration for this purpose, he needed to come up with a second insight."

"The warping of space and time?"

"Correct. He set about proving that space-time actually warps, or curves whenever mass/energy is present."

"So space-time warps whenever energy is present as well?"

"Correct."

"So if there is a body, such as a star, space-time will curve around it?"

"Yes. In the direction of the body. In a sense, this is how the force of gravity is expressed."

"So space-time is not only one continuum, but is a continuum that warps in the vicinity of any mass/energy, and in the direction of that mass/energy?"

"Affirmative."

I take a long sip from my mug, watching him do the same. I push away from the table, hold the mug in my lap. "This warping of space-time, if I'm correct, was confirmed when the behavior of light from stars was studied during a solar eclipse."

"Indeed. The whole thing's quite amazing actually. A solar eclipse, Einstein believed, provided an opportunity to obtain experimental verification. Light from stars is like tiny, far-off pinpoints, he insisted, pinpoints which we see at night, but not during the day because of the overwhelming light of the sun. But during a solar eclipse, when the moon blocks the sunlight, these stars become visible. At this point, we can witness just the kind of effect the sun has on these beams of starlight, since these beams must still pass close to the sun during their trek to Earth. As Einstein's general theory of relativity predicted, the sun indeed alters the path of the starlight."

"Because the sun causes the surrounding space-time to warp," I offer, "the light's path is bent by this warping."

"Correct. And the bending of the path is the greatest for those beams of light that pass the *closest* to the sun. Again, according to Einstein, this is how the force of gravity is expressed."

"Didn't this historic eclipse take place somewhere around 1920?"

"It was May 29, 1919 to be exact," he answered. "This was the solar eclipse that verified Einstein's prediction based upon general relativity. A verification that overthrew all of our previously held conceptions regarding space-time."

"It's one thing ... and it's malleable."

"Yes. And what's more Einstein also proved that acceleration warps both space and time." He smiles. "He basically proposed that all gravity itself really is, in the end, is the warping of space and time."

The Bridge

I sit down, but push away from the table, lean back and clasp my fingers together under my chin, thoroughly thinking through what he had just said about gravity.

He continues, "Remember how Newton thought that gravity was some instantaneous force that mysteriously tied one body to another? And if something happened to one of the bodies, this tie was broken, and the attached body felt an instant effect?"

I nod. "And it was this instant effect that troubled Einstein."

"Correct. What Einstein proposed was that space is flat and somewhat like a fabric. Picture a hammock that someone may have tied to two trees in their three acre backyard, but this hammock is not made out of netting, but of a smoother, equally flexible surface. Now picture this hammock hundreds of times bigger, and imagine further that it's no longer spread just between two nearby trees, but is spread out all over the entire yard." He pauses for a moment, thinks something through. "Though this is only a two dimensional image," he nods, obviously answering his own question, "it will still provide a clear illustration. Now," he turns his words back to me, "if you can, picture placing a cement-filled basketball anywhere on that hammock. What will happen?"

"It would sink ... causing the hammock, in essence, to warp around it."

"Correct. Now let's say that a tiny marble had already been placed on this hammock on the other side of the yard. When the basketball was set in the net, what do you think happened in regard to the marble?"

I think for a moment. "It would no doubt feel sort of a ripple effect, almost like a slight vibration."

"Would it move?"

"It would definitely move ... in the direction of the basketball."

"Now envision that the basketball is the sun and the marble is the Earth. The marble moves toward the basketball, seemingly attracted to it. And as it approaches, does its speed increase or decrease?"

"It increases."

"So the *shape* of the fabric, of space, determines both the direction and the speed, essentially the time, of the marble."

I nod, seeing it becoming more clear. "This supports what Einstein had found with time and space being linked."

"Correct. The link between accelerated motion and gravity is tied in also with the structure of space."

"Which is expressed in the curving of space around anything with mass or energy."

"Such as the basketball, our sun, and the marble, our Earth. Now," he holds up a hand.

"What happens when the marble reaches the basketball?"

"With the right speed now behind it, the marble would have to move in a curved path around the crest, or the mouth, of the actual warped area where the basketball exists." I nod, seeing it more clearly. "It would, in essence, go in orbit around the basketball."

"Which explains, in a very basic way, how and why our Earth is in orbit around the sun."

"So," I push forward, "obviously our Earth is also, though to a much smaller extent than the sun, warping the fabric of space-time, causing its own effect on bodies smaller, or with less mass, than itself."

"Like the moon?"

I nod. "Yes, like the moon." Then I look up at him very closely. "And on and on it must go, orbiting body upon orbiting body … like some great linking chain, or an unending malleable net with millions of linking meshes."

"Indeed. And remember, the sun itself is in orbit as well, so it too is just one of those millions of links … or meshes."

"This apparently resolves Einstein's dilemma with the instantaneous reaction proposed in Newton's theory," I conclude, "A mass causes a ripple-like effect in the fabric of space-time, but not an instantaneous effect. The example of the basketball and marble illustrates this. The marble feels the oscillations from the basketball being present, but not instantaneously."

"Exactly. And just think if there was no large mass present—like the basketball in our scenario—on the fabric of space-time, the marble would just remain at rest or move along at constant speed until a force acted upon it. A force such as the one created by the basketball. At that time, space-time would be warped, the fabric would oscillate and the marble would soon feel its effect."

"Has the speed of these traveling oscillations ever been determined?"

"Einstein concluded that they travel across the fabric of space-time at exactly the speed of light."

"Not instantaneously?"

"By no means. Nothing moves faster than light, not even gravity."

I move closer, resting my elbows on the tabletop. "What happens if the composition of the sun, or another star, changes altogether."

He studies me. "You mean like if a star began to die?"

I nod. "Can you explain?"

He sits back, stretches his legs out before him. "Some stars are too large to sustain their weight and eventually actually begin to collapse in on themselves."

"So their size begins to dwindle?"

"Yes, but not their density."

"Their density increases?" I ask.

"Absolutely. And this first stage of a star's death, or collapse, makes it what is called a *neutron star.*"

"So they become small with high density?"

He nods. "Picture all the mass of our sun collapsing into an area no bigger than thirteen miles in diameter. That's how small and dense they can become. They, in essence, become a bundle of tightly packed subatomic particles."

"And once they reach a certain point of collapse, a certain critical radius, don't they become a black hole?"

He nods emphatically. "And this certain point of collapse, as you call it, this certain critical radius is called the Schwarzschild Radius, named after the German astronomer Karl Schwarzschild. He was the first to propose that if a star collapses to this certain critical radius, then the warping of space-time would be so great that anything, including light would not be able to break free. They were called black holes just for this reason—that even light cannot escape them."

"So they emit no light, even the light that has entered them?"

"Correct."

"Now if a space object travels near a black hole, how close does it have to be to become trapped in its grasp?"

"That depends on the size of the black hole, and the distance the object passes by it. The area along the rim of the black hole that determines the line that cannot be crossed by passing objects is called the event horizon. Once entering the event horizon, an object finds itself suddenly approaching the speed of light as it is sucked toward the center of the black hole. At this point, it's too late—the object will soon be lost forever."

"And the closer it moves toward the center, the stronger the gravitational pull?"

"Absolutely."

"So as the star grows smaller and smaller, then the center must grow smaller and smaller ... then what takes place at the heart of a black hole?"

"Though no one knows completely for sure yet, the prevailing theory leads us back to our parallel universes discussion."

I watch him as he leans onto the tabletop. "See, if enough matter is collected in one place, as at the heart of a black hole, then the fabric of space-time, according to some scientists, would be so distorted that it would actually rip."

"So the laws of physics, at this point, would completely break down?"

He nods. "By the time you reached the heart of a black hole, your mass would be reduced to zero, and you would find yourself in a situation where time and space no longer existed."

"A singularity?"

"Essentially, yes."

"But singularities aren't supposed to exist, right? In the past, weren't theories that ended in a singularity, or an infinity, thought to be flawed?"

"At one time, yes. But that may not now be true. Singularities, infinities, may exist after all, and in fact may be a vital feature of our universe."

"You mentioned parallel universes as the prevalent theory."

"Indeed. When this center point is reached and the fabric of space-time is indeed ripped, many scientists are now postulating that another universe, at that precise moment, is entered. This crossover point from one universe to the next is the *Einstein-Rosen Bridge*, named after Einstein and an associate physicist he worked with at Princeton named Nathan Rosen."

"The crossover point between one universe and another," I repeat.

He nods. "The bridge ..."

I think for a moment. "A thought just jumped into my head."

"What is it?"

"Aren't virtually all celestial bodies in motion?" I ask.

He looks at me carefully. "It appears indeed they are."

"Then in this case, the black hole itself would not be static as it decays, but would still be spinning."

"Neutron stars that are rotating are called pulsars," he adds.

"How would this fact—that they are spinning—play into the parallel universe theory?"

He nods with a smile. "Excellent question and one that has been posed before. In the case of a spinning black hole, scientists believe that the end point would lead not just to another universe, but to an *infinite* number of universes."

I stand up and again begin to pace. "I find myself returning to the quantum realm," I tell him, "and most specifically the double-slit experiment where, according to Feynman, the electrons apparently traversed every possible path through the slit before settling on the screen."

"Sum-over-paths."

"I mean, what we just discussed—you know infinite universes and all—would normally seem far-out and outlandish, but it doesn't because it seems consistent with what the quantum realm has thus far shown us."

"What is true at the micro level seems to also be true at the macro level," he reminds me. "After all, the entire universe may have begun within a single point of energy no bigger than one one-hundredths of a grain of sand."

I stop and look down at him. "If it turns out that an infinite number of parallel universes do in fact exist in the micro realm, then what does that hold for us in the realm in which we live and experience?"

"That's a logical next question. In fact, it's one that's already on the table, particularly when paired up with what the quantum realm is indicating to us. If it's indeed the case that when single particles in the double-slit experiment are presented with a choice, they seem to branch off, quite possibly into parallel universes to traverse every possible path, then the question remains: when we are faced with a choice on our own sensory level that manifests itself in some physical manner, do we somehow not branch off into parallel universes in order to traverse through every possible outcome of our decision?"

"All the while, we would only *experience, or be aware of,* one universe at a time, as in the way the photon in the double-slit experiment seemingly conspires to show us only one of the outcomes of its multiple journey."

"These are theories still on the table in the world of physics."

Now both exhausted and invigorated at the same time, I push to my feet, shuffle over to the window, search for rabbit and fox tracks in the snow. "What does this all hold for us?" I repeat with a sigh, seeing what looks like only a single trail leading into the underbrush of the forest, "what does this all hold for us?"

9

Rudy looked up to see Sinclair standing over him. "You still reading, Dad?"

Rudy nodded. "Just one more chapter …"

"What time are we gonna leave?"

"We'll hang out for another hour or so," Rudy said, "and then we'll head out. Sound good?"

The boy nodded. "Sounds good."

Parallax

I settle in again across from him, bright rays of snow-reflected sun angles sharply across the room. The light is so white and brilliant as to seem surreal.

He watches me curiously. "What's on your mind today?"

I shake my head with a smile. "All the possibilities."

He nods. "Yes. And there are many."

"I don't know where to begin."

"Begin anywhere," he says.

"Okay," I sit back, look directly at him, "the double-slit experiment."

"Alright," he leans forward, rests his weight on his elbows.

"When the light is dimmed to where there's only one photon approaching the screen where *both* slits are open," I begin, "something strange apparently occurs when that photon is presented with the choice between these two slits."

"Apparently so."

"When it hits the photographic plate behind the screen, it behaves as if it is interacting with other photons." I reestablish.

He nods. "Instead of hitting the plate as you would think a single photon would—as a single photon, in fact, does when only *one* slit is open …"

"As a single band on the screen."

"Right. But instead," he holds up a finger, "it creates an interference pattern when both slits are open."

"Which is what is created when a full beam of light with *more* than one photon is directed at the screen."

"Correct."

"And that leads us to the question of how?"

"Yes."

"Along with the possibility that the particle is simply responding in a discerning way to the situation, the other theory on the table is that it interacts with another photon from an adjacent, or closely aligned universe. Or," I hold up a hand to correct myself, "a closely aligned relative state."

"That's right. See, the experiment can be performed in two ways. One way places detectors next to the slits to actually observe the photons as each passes. When these detectors are in place, the photons behave as particles, going through one slit or the other. They hit the screen in the anticipated way, creating a single beam that corresponds to the chosen slit. But when the detectors are taken away, the craziness begins. The photons now, as amazing as it is, create an interference pattern."

"Behaving now as waves."

"Correct. As if, according to Feynman, each particle is now going through both slits simultaneously."

"Maybe there are, in fact, two photons, one of which we cannot detect. After all, the behavior clearly indicates that there are two photons. This is the reasoning, correct?"

"Correct. It is precisely this point that led to David Deutsch's multiverse theory … our multistate."

"That another photon is overlapping from another r-state?"

"Exactly."

"One that we cannot detect in any direct manner?"

"Other than for the interference pattern that is created because of its presence."

"Let me ask you—if this is true, what is the other r-state experiencing?"

"I would imagine the same thing we are … except they are seeing the mysterious interference pattern caused by our photon, invisible to them, overlapping into their r-state, living out the other possibility apparently not chosen by their photon."

I push out of my seat to begin my ritualistic pacing. "There's a thousand directions we could go at this point," I contemplate as I move in front of the window to stare out at the quickly melting snow. "Amazing and beautiful possibilities." I turn back, lean against the wall, tuck my hands deep into my pockets. "We could talk about the observer/measurer aspect for hours, but let's stay with this relative state possibility for a while." I sit down across from him. "This strange behavior we've been discussing also holds true for electrons. Correct?" I ask.

He nods. "And apparently for all other particles as well."

"According to the theory, whenever presented with a choice, these particles actually branch off into other r-states and live out all the possibilities of that choice?"

"In essence—yes. Let's take the double-slit experiment a step or two further. David Deutsch does a great job explaining this experiment in his book, *The Fabric of Reality*. I only hope my abbreviated version does it justice." He adjusts in his seat as if to ready himself. "Okay," he begins, "let's say we have a beam of light that encounters our same double-slit barrier. The slits are, in this case, one-fifth of a millimeter apart. We essentially know what will take place when this beam comes across these slits."

"It will create an interference pattern."

"Correct. We would see those bands of light and dark on the screen. To be precise, if the screen was three meters from the barrier with the slits, we would see five bands of light upon the background of dark. "Now," he holds up a finger, "if we cut two more identical slits in the barrier, so that there are four total slits, what would we see?"

"How far apart are the slits from one another?" I ask.

"One-tenth of a millimeter."

"I would think that four slits would create more light bands," I answer, "and maybe the bands would be a bit more blurry."

"Your conclusion is perfectly logical, but it is incorrect. What happens is that an even more complicated pattern is created. A pattern with surprisingly fewer bands of light and dark, instead of more. Three bands of light are created on the backdrop of dark instead of five. Just the opposite of what one might surmise. The areas of darkness seem to grow larger in-between the bands of light. Obviously the light from the added slits has interfered with the light from the original two slits, causing bands of light to show in different places with four slits open than it did with just two slits open."

"So places that are dark on the four-slit experiment are illuminated on the two-slit experiment?"

"Essentially yes."

"Just like those waves of water we talked about earlier merging together after passing through the slits, thus interfering with each other's path to the screen. The interference that's taking place with this light would not happen if only one slit were open."

"But happens," he adds, "when both slits are open, and happens even more when four slits are open. Something is clearly interfering with the light to create the interference. And that something is *light*. That's the key point."

I study him curiously. "So I guess our next step in this experiment would be to see what happens when the light is dimmed so only one photon at a time is fired at these four slits?"

He smiles. "Correct again. But before we do, what do you think happens when an obstruction of some sort is placed in the slits, an opaque obstruction that effectively blocks light?"

"If one slit is obstructed as opposed to two, then we see just a single beam of light hitting the screen as one dot. If two slits are obstructed and two open, then we see the two-slit open result we just described a minute ago."

"That's all correct. And if we put in an obstruction which the light was able to penetrate, that was *not* opaque, what would happen?"

I think for a moment. "Well, if the light was able to penetrate it, then the result would be the same as if there were no obstruction."

"Right. The key is if the light can pass through it. If it can, then the interference pattern will happen as we have seen. The essential point is that the interfering agent is light, nothing else."

"And if only one photon approaches the four slits?" I ask again.

"You tell me."

"Well, if we accept that what's causing the interference is other photons, then common sense would tell us that the interference pattern would not be created. How could it be if there are no other photons but the one?" I press on. "I would think we would see a single dot on the screen behind whatever slit the photon chooses to go through." Then I stare at him. "But I have a feeling this is not what happens."

He nods. "This is not what happens."

"An interference pattern is still created?" I ask.

He nods again, silently. "If two slits are open, we see the bands of light and dark. If four slits are open, we see entirely different band locations of light and dark."

I shake my head in disbelief. "Just like before."

"Just like before."

"Let me play devil's advocate and ask the obvious questions. Does the photon interfere with itself somehow?"

"Well, here again, if we put detectors on all the slits, only one photon passes through one slit at any one time."

"Does the photon split, and then the parts interfere with each other as they come back together?"

"Again, the detectors only detect one photon going through one slit at a time. Only one detector goes off." He shakes his head definitively. "No, it seemingly doesn't split in any way."

"Then what's interfering?"

"Well, we've determined that it is light that interferes with light. Again, by placing obstructions that block the light proves this. When the light was blocked from going through the various slits, the light going through the unobstructed slits behaved in the predicted way—a way that happens when it's *not* being interfered with. So, again, we know it's *light* that interferes with light."

I laugh. "But there's only *one* photon! The obvious question that one will ask once again is: if there's another photon involved, where is it and where did it come from?"

"The answers could quite possibly lie with what Deutsch calls shadow photons. I've always referred to them as *mirror* photons."

"Mirror photons?"

"Identical photons located in a closely aligned r-state, which are only detected in our world by the interference pattern their presence creates." He shrugs. "As amazing and bizarre as it is, this explains what's happening ... it's a viable theory to explain this single photon interference phenomenon."

"These mirror photons are obviously invisible to us in our r-state," I add.

"And undetectable. Again except for," he holds up another finger for emphasis, "being detected indirectly through the interference they create when they come in contact with the photons located in our r-state."

"These photons that are located in our r-state are tangible in our world?"

"Correct. But they are apparently able to be affected by their counterpart found in the closely aligned r-state. See, when one of our r-state photons encounters a choice and chooses one path, as in the various slit options in our experiment, a vast number of mirror photons pass through the other slit options and cause the interference."

"Thus the interference pattern."

He nods. "Thus the interference pattern."

"How many mirror photons are there for each of our photons?"

"No one knows for sure, but there seems to be a significant number for every one of our r-state photons. Deutsch believes about a trillion or so."

"Are photons the only particle to have mirror counterparts?"

"According to the evidence, the answer seems to be a resounding no. All particles have mirror counterparts, and they all are capable of interfering with one another. Electrons have mirror electrons, protons have mirror protons, neutrons have mirror neutrons, and on and an on it goes ..."

"How do we know for sure?"

"Well, our r-state, *this* physical universe, is made up of particles which are concrete in this r-state of ours. They come in contact with and interact with each other, and are measurable and observable by our bevy of measuring devises, including to a certain degree our sensory apparatuses."

"Our five senses."

"Yes—our five senses experience them to a limited degree."

"Now, due to the interference phenomenon that we've been talking so much about, our r-state particles come in contact with particles from the rest of the multistate, which is a much larger reality than what we are capable of experiencing directly."

"Correct."

"And are all of these mirror particles located in this one large 'other' reality? Or are they separate from one another in many realities?"

"From what the evidence and experiment indicates—there are many r-states, all separate from one another. Again, evidence indicates that concrete particles in one r-state are the mirror particles in another r-state. And the different r-states are only detectible by one another in the same indirect manner—through the interference phenomena created by the overlapping particles. And everything's the same in regard to the physical laws that each r-state must obey. The only thing that's really different are the locations of the particles."

"Okay," I hold up a hand of my own. "Let's slow down here. So all of these different realities, these many different r-states, constitute the multistate we've been alluding to?"

He nods.

"But still, how can we conclude that other r-state particles, besides photons, have identical particles in other r-states?"

"The best example is to ask the question: if there are these mirror photons that are apparently blocked by an opaque obstruction that we construct in our r-state, then are they blocked by the atoms of the obstruction?"

"Okay," I nod, as if to say *continue.*

"Again, we go back to experimental evidence. When we say the obstruction blocks light, we're basically saying that it is absorbing the light. When light is absorbed, the change in energy is detected in the atoms that absorb it—in this case, the atoms of the obstruction. When we set up an experiment to block our r-state photons that were being interfered with by mirror photons, we see that only *our* r-state photons are absorbed by the obstruction's atoms. The mirror photons that we've concluded exist because of the interference phenomenon do not register as being absorbed by the atoms of our obstruction when it is in place. The state of the atoms is not altered in any way by the mirror photons. Basically, photons from another r-state do not intermingle with *atoms* from our r-state. They only are capable of intermingling with, and subsequently affecting, their like particle, which in this case are photons."

"But the particles, both from our r-state and the other r-state, are affected when hitting the obstruction?"

"Correct, they are affected by the obstruction in the same way—in essence, both are stopped. But again, the *obstruction* is not affected by them in the same way. It is only affected by our r-state particle."

"Then the question still dangles—what's stopping the photon from the other r-state? If atoms from our r-state don't seemingly interact with photons from another r-state, then how are these other photons halted?"

"Perhaps by a mirror obstruction closely aligned to our obstruction. What Deutsch refers to as a 'shadow barrier.'"

I inch my chair away from the table, stretch my legs out before me, and cross my arms over my chest. "Do you know what's suggested by that?"

He nods. "It suggests that our obstruction has a mirror obstruction, and that the atoms of this mirror obstruction are in fact able to interact with *its* photons."

"If this is the case, it's able then to absorb the photons from its own r-state, thus causing the apparent halting of these photons when our obstruction is in place."

"That's the way I see it."

"And so," I continue, "any measurements taken in that r-state would show those atoms gaining energy as they absorbed *their* photons."

"And our photons which, remember, would be causing the interference pattern in their r-state, do not register with the atoms of their obstruction."

"Everything's the same for them, except sort of reversed."

"That's why I use the term *mirror.*"

I stand up, and begin my pacing. "It's all so bizarre and outlandish."

"It's the best explanation I've heard to date. I mean, I'm open to another explanation. But I haven't heard one as of yet that explains as well what's happening regarding the interference phenomenon, and all that goes along with it. One thing I can tell you with some surety though—whatever explanation winds up being correct, it will be just as wild, bizarre, and seemingly outlandish. Prepare yourself."

"So," I continue right in stride, "if the obstruction, the mirror obstruction, the photons, and the mirror photons in our illustration are part of this structure of reality, so must all other physical entities be a part of this structure of reality. If all the atoms that make up our obstruction have mirror counterparts, then all physical entities have mirror counterparts. If," I hear myself pretty much rephrasing the same thought, "all the atoms of the obstruction have identical mirror atoms, and these mirror atoms make up an identical obstruction in another r-state, then do all physical bodies have the same set of circumstances that constitute their own reality?"

"It would stand to reason that they do."

"So this table," I lean down and tap the smooth wood, "has a mirror table that has real existence in another r-state?"

"It would seem so."

I pause, stare at him. "And then there has to be another mirror of all the atoms and molecules that constitute you. Therefore, another 'you' exists in all of these r-states."

He smiles. "And another you."

I shake my head, make my way over to the window. "That's downright spooky."

"As spooky, as bizarre, as outrageous as some of the beliefs we, as a species, have been living under for most of our existence?"

I nod, and seeing the endless blue sky through the frost of the thin glass, acknowledge the truth of his words.

"One thing's for sure, what we see, what we experience, is not all there is to reality, to truth."

I return to my chair, lean my elbows on the table and feel the sun warm my face. "This leads me back to my final question from the last time I was here: what does this hold for us in our lives, in our level of life and experience?"

He sits back, crosses his legs. "Let's look at it for a moment. If every choice that leads to a *physical action*," he pauses, holds up a finger for emphasis, "I must add that little caveat—in my view, it needs to lead to a physical action. Anyway, if every choice that leads to a physical action causes a branching off into multiple r-states to live out all the possibilities of that choice and its subsequent action, then the implications for us are indeed far-reaching."

"Let's just take the scenario," I suggest, "that you're walking down the street and you can make either a left or a right at the end of the street to get to your destination. Either path can get you where you want to go just as easily as the other. Let's say that when you get to the end of the street, you happen to go left, where you could just have easily gone right. By going left, according to the theory, you are not only living out, in this r-state, the outcome of having gone left, but you are also, in a separate r-state, living out all the possibilities of having gone right."

He nods.

"Now, let's break this down. What happens if you're not careful when you turn left and you step out into the street and are struck and killed by a truck? What does this scenario hold?"

"Does the fact change that the choice has been made?" he returns a question.

I think for a moment. "No."

He shrugs, studies me curiously.

"Then we are still living out all the possibilities in another r-state of the other side of the choice, which is turning right?"

"It would appear as such."

"Then we are dead in one r-state, but still alive in another?"

"So it seems."

I sit back. "That's incredible."

"How's this," he uncrosses his legs to sit forward, "what happens if, as you're walking down the sidewalk before you actually make your choice, a woman comes around the corner and bumps into you, thus making it easier, because of the angle that she bumps you, for you just to turn right? Though, don't forget, the choice to turn left still exists, but now at a lower probability."

I think through the scenario for a moment. "You would live out," I answer carefully, "in this universe, this r-state, the possibilities of turning right, while still living out the possibilities of turning left in another r-state."

"What happens then to the woman that bumped into you? By you having been there when she made that turn, she now, because of the collision, turns left at her next turn instead of right. Let's say you had dirtied her shoe and she decided to turn left to get a napkin at the convenience store instead of turning right toward her original destination. What does this entire interplay mean?"

"Well, obviously she's now branching off to experience all the possibilities of both turning left and right. Left in our universe, right in another universe."

"What does it mean that one being, or physical entity, affects the other?"

"Well, I guess," I answer slowly, "though it complicates and seems to muddle the entire picture, it makes sense, particularly when you consider the ultimate theory of the unified field."

"True," he nods in agreement, "we can never forget that. All these theories may in fact be just features of the unified field. And if you think about it," he adds, "in the double-slit experiment itself, even the photon is affected because of our presence. In this instance, the photon had the choice between the two slits because we structured the experiment in such a way to produce only one photon moving toward the screen. We can never forget about everything existing within the framework of the unified field."

I push my chair away from the table. "Okay, let's pause for a moment," I suggest with a laugh, "we're starting to move too quickly. Let's back up a bit."

"Agreed."

"We could go on ad infinitum with various scenarios in regard to the multi-state. But one thing in particular comes to mind—are the possibilities inherent in every choice infinite?"

"Hmm." He sits back again, props his elbows onto the arms of the chair, presses all five fingertips together on each side of his chin. "Well, if you made a right at the corner, and there was nothing else there ever again on that path to interact with you, I would say it would be finite."

"It seems unlikely though that a situation would occur where there is nothing else ... no choices later for a decision of possible directions to be taken, no other people or other entities that could cause you to make decisions along the way, no change in weather, etc ..."

"Unlikely indeed. The only way you could not face a choice would be to find yourself in a vacuum of some sort where nothing would act upon you...but that's another discussion altogether."

"Suddenly this is reminiscent of the principles behind Newton's laws of motion," I grin.

"Like I said—another discussion," he lets out a deep breath, ponders. "But where we are in our discussion here, I would have to say that what we experience within the realm of our separate selves is finite, but the manifestations of our choices carry on infinitely. The ripple effect of our choices impacts infinitely, as do those of the photon, the electron, and the like ..."

"Sounds a bit like the *chaos theory*."

"It does. Just as you bumping into that woman helps determine her direction, she interacts with and thus determines outcomes for an infinite number of scenarios on that path, not to mention the infinite number of scenarios on the path not taken that are being lived out in the multistate. Maybe when you stepped on her foot, you not only dirtied her shoes but you caused her to twist her ankle a bit. That would have an effect ..." He raises his hand up in the air, "let's say, for instance, she became quite annoyed by the whole occurrence and it changed her mood. She finally makes it to work and immediately yells at a coworker. Now that person's affected ... on and on it goes."

"But what you said a moment ago that what *we* ourselves experience, even in all the r-states combined, is finite? Why?"

"Because of our physical existence as human beings, we can only live a certain amount of time, within this r-state and within the other r-states. Just as we brought up the scenario of turning left at the end of the street and being struck and killed by the truck. In that situation, we can see how you could be both alive and dead because there was still some natural life to be lived. Dead in this r-state, but still alive in another. But what about natural death? Does someone who dies from natural causes branch off to remain alive in another universe?"

I stare down at the tabletop, think it all through. "I guess not," I finally say. "If these other r-states are closely aligned with our own, then the laws of human longevity will have to hold true there too."

"Precisely."

"So once our natural longevity is up, it ends?"

"For us ... most likely, yes." He smiles. "You know, to continue on in this direction of our discussion, we'd have to begin contemplating questions of a more metaphysical nature."

I nod, but push ahead undaunted. "Once our physical existence ends, do we continue on in any way?"

He smiles. "That's one of the primary questions, isn't it?" He runs his fingertips along the length of his beard. "My rule always is: if I can't come up with at least a plausible physical theory for how something could take place, I discard it until I can."

"What about the law of the conservation of energy?" I ask. "Does it possibly open up a door for such a thing?"

"You mean the law that says energy is neither created nor destroyed, but only changes forms?"

I nod.

He pulls in a deep breath. "To a degree, it does." He looks out toward the window, thinks for a moment. "I mean, who knows what the energy takes with it as it changes and moves on ..."

"Does it take a piece of what we perceive as being 'you'?" I ask tentatively. "But not in the traditional sense of the word," I add quickly. "Maybe," I hear myself choosing my words carefully, "a remnant of the consciousness that was 'you' becomes engrained in the energy in a certain way … creates a kind of oscillating marker of sorts …"

He smiles. "Sheer speculation, you know?"

I nod. "I know, but how do we know that what became 'us' was not already present within the energy, and thus came forth as one in a long line of manifestations of this energy …"

He stares at me curiously. "Incarnations that come into existence at various points, in various forms …" He pauses. "Interesting indeed, but at this juncture, sheer speculation."

I sit back with a sigh. "I know."

He leans toward me, rests his hands on the table, speaks very quietly. "In my mind, the best chance to come across the knowledge of such things, if they do exist, is through the study of the natural world." He taps the tabletop for emphasis. "This study we're engaged in … of exploring the possibilities shown to us by Nature." He pushes his chair away from the table, extends his legs out in front of him, rests his hands in his lap. "As much of a realist as I think I am, I've admittedly gotten to the point now, with as wild as some of these new theories are getting, not to rule *anything* out. Though I think that if we ever do find something like you put forth—you know, a piece of us, be it the soul or whatever, continuing on—it will not be what we always thought it would be." He now gestures with both of his hands. "It will probably be something even more fantastic than we've ever conceived of before. And it will work and be *explainable* on the physical realm, which will give it validity. But it will take hard work, study, and long periods of living amidst uncertainty to obtain. Everything else, before the hard work is done may just be sheer speculation, largely based on fantasy and myth. But if we're willing to ride our natural curiosity, stand strong against uncertainty, explore, question and evolve our understanding of Nature, *as Nature is*, we may stand a chance of finding what so many seek …"

I nod. "What you said just a few moments ago reminded me of something I meant to tell you. And it sort of relates to what we're talking about. You remember during one of our previous conversations I mentioned that I sometimes awaken suddenly in the middle of the night with a thought that I just have to write down. Well it happened last night, and I felt my way around in the dark to my desk and wrote it down." I pull a piece of folded paper from my pocket and slide it across the table to him. He picks it up, unfolds it, reads it aloud. *"Remove yourself from the equation, and all will be clear."* He looks up at me with a nod. "Hmm."

"By the time I got to my desk, though, I had lost the meaning of it, but I thought it was interesting."

He hands it back. "Quite."

I shove the paper back in my pocket. "And what you just said a minute ago I think hints at it. You said: 'what we experience, within the realm of our *separate selves*, is finite'." I nod with a definitive tap of my finger on the table. "That could be the key. When we view, and realize ourselves as being separate from the harmony of Nature, we are then trapped in a finite existence. But once we truly—and like you said, it takes work, patience, and long periods of uncertainty—but once we *truly* recognize ourselves as being part of this unified whole, with no separation from it, then we are no longer trapped within a finite existence."

"And it may begin with the simple recognition of the unbelievable harmony of Nature. Once you really feel that harmony working around you, and within you, it begins to release you."

"Isn't that very close to Einstein's 'cosmic religion', as he called it?"

"I guess it is," he nods, "Einstein loved the Dutch philosopher Benedict Spinoza, who in essence believed that 'God' could be found in the study of Nature, and that the two were indistinguishable."

"Basically, according to Spinoza—and to Einstein himself—the greater an individual's understanding of Nature and the universe, the nearer that individual comes to 'God'."

With that, we both find ourselves quiet for a few moments. Then he smiles. "I have an idea. Even though there's snow on the ground, it's pretty warm out. You wanna take a walk?"

"Great idea."

◆   ◆   ◆

A few minutes later, we find ourselves standing outside in receding, but still bright and brilliant sunshine. The top of the snow, now slowly melting, stretches before us like so many tiny granules of shimmering diamonds, and the air is crisp and glorious. We walk slowly away from the small wooden dwelling, and in a few moments I find myself standing in almost the exact same position I had just a day or two earlier when I had watched that fascinating chase between the rabbit and the fox. Still able to see a vague trace of their paths, I point it out to him. "That's where I saw the fox take off after the rabbit."

He nods. "If you will remember, that's what led us into the whole discussion of Schrodinger's cat, and eventually into the parallel universe theory ... the relative state theory."

"You know, I think the whole relative state theory appeals to me so much because it fits in with a natural belief, almost like a natural instinct, I had as a little kid. I always sensed as a child, for as long as I can remember, that we all are going to live each other's experiences. As I grew, it seemed like such a wild, far-fetched, almost metaphysical theory, but yet I never fully discarded it as I normally would simply because I seemed to be born with it. As crazy and as implausible as it was to me, I never let it go because it came so naturally. It seemed to have been a part of my genetic package, if you will." I tuck my scarf inside the collar of my coat. "Now comes the relative state theory, which seems to work in regard to that natural childhood theory. In essence, the multistate allows the possibility of living all experiences, or nearly all experiences, plausible."

"Yes it does," he nods, his hands shoved deep into his pockets. "It certainly does."

"And to add on to what we talked about a few moments ago regarding the conservation of energy—if we can ultimately come to identify as much with energy as we do with ourselves, with this sense we all have of a separate self, or," a second thought occurs to me, "better yet, if we actually come to the full realization that the energy is indeed us, that we are not separate from it, then our entire world view would have to be altered."

"And the unified field theory," he offers, "to me at least, hints at just that."

"And the more I think about those two possibilities that could explain what was happening with that one photon during the double-slit experiment—you know, that the photon goes through both slits at the same time, or the whole multistate explanation—both possibilities can absolutely revolutionize our lives."

"How so?"

"Well, if we take a look at the entire question concerning the observer/measurer, the ongoing Einstein/Bohr debates were all about this question. The idea of continuity versus discontinuity."

"The idea that the observer/measurer indeed plays a role in determining reality," I add. "In the case of the double-slit experiment, it's all about how we, as the observer/measurer arrange the experiment."

"So what you're basically saying," he says, "is that these things add value and credence to the role of the observer/measurer in overall reality."

"Precisely. And if this all is indeed true, then we may need to rethink ourselves, and our role within the world."

"In what ways?"

"Well if the observer indeed plays a role in determining reality, in affecting the natural world, then at what point, if any, would our own observations begin to make a mark on the natural world itself?"

"One individual?"

I think for a moment before shaking my head. "I'm afraid that one individual, or a group of isolated individuals, just wouldn't be enough. But maybe if many people began in earnest to study, to *observe* Nature at a more intense level, then maybe that would, at some point many generations later, begin to have an effect on it."

"By many people, do you mean a critical mass of people?"

"You could put it that way."

He nods as we walk. "Let's approach it this way. Do we affect Nature in any way?"

"In a sense, I suppose Nature's continually adapting to us almost as much as we are to it."

"In what ways?"

"Well, for one, it's constantly trying to regenerate itself after we take from it. And the more we take, the more it has to regenerate."

"Which, by the way, it's having a more and more difficult time doing," he adds. "It's beginning to try, I'm afraid, to shed us," then he waves his hand, "but that too is another discussion." He takes a deep breath to regroup. "Let me ask you this: if we were no longer here, how would the natural world, the universe respond, if at all?"

"I imagine, though only momentarily, it would have to adapt to our absence as well. I imagine it simply sidestepping and continuing on."

"Conversely, like you said, what would be the cause and effect scenario if we, as a mass, began to study the natural world at a more intense level? What would be the first manifestation of adaptation that would stem from this?"

I nod, remembering what he had proposed during one of our earlier discussions. "Thinking back to what you said your first step would be. You remember?"

He nods with a smile. "Ten minutes a day of study, followed by ten minutes of meditation?"

"At some point, something about Nature would capture the imagination of just about everyone, and in some of those individuals, lights would start going on. Imaginations would be sparked. Questions would start being asked. The occasional epiphany would occur."

"One would hope," he says. "But let me ask you, what practically would begin to happen if this did begin to occur?"

"Well, I imagine that those who were indeed moved would begin to view the totality of Nature differently. And this could, I can only suppose, lead to a greater respect for Nature. With this being the case, then how those people treated Nature, how they valued the world would change, right down to how we implement our technological advances." I shrug, and notice for the first time my feet wading through the snow. "And what you said—how Nature may be trying to shed us at this point—may be altered because of this change in us. If our view of the natural world changed, followed by a greater outward respect for it, then maybe it would have less of a need to shed us. If we lived more in harmony with it, then maybe it would keep us a bit longer while it progressed along on its own evolutionary path."

He pauses, nods as he looks at me. "If Nature is in fact continually adapting to us, then it would surely adapt to this change in us as well." He begins to walk again. "See, my fear is that this species has lost such touch with the true beauties of life that we're forcing the world to guide us into our own extinction. In our rampant greed and desires, we are constantly taking but yet contributing very little back. Just like what you mentioned about technology … if true change occurs within us, then the ends toward which we project our technological advances and breakthroughs would surely change as well. If we're not contributing to our own evolution, then how can we contribute to Nature's evolution? But this new respect and understanding of Nature and the universe may in fact actually help to keep the species from its own extinction, help it to evolve within itself as well as evolve into its rightful place within the whole."

I nod. "It very well could." Then I smile with a final shrug. "So that, from my perspective, would be the first manifestation of a heightened observation."

"But would our observations *in and of themselves* begin to have an impact on the natural world?"

I take a deep breath and think it through. "Our only chance would be not as individuals, but as that collective mass that we've been talking about. Apparently, Nature presents itself to our species in certain ways. Such as the double-slit experiment. Maybe light presents itself to us in a way we can understand. When a beam of light is projected when both slits are open, our own eyes see only two beams hitting the screen. We do not see what is actually happening, which the photographic plate picks up. Maybe we are shown only what we're capable of seeing." I shrug. "I don't know, but maybe if our overall understanding *as a species* became greater, then maybe Nature would begin to present itself to us differently … in a way that coincides with our new level of understanding. You never know, but this could be the next challenge we face that will determine if we evolve to the next level or become extinct. After all, if Spinoza's correct, then the more we'd be seeing of Nature, the more we'd be seeing of God, of truth."

He looks up at me. "Now we're back to your consciousness question."

I shrug, pausing in the snow. "I'm like you, I can't rule anything out at this point. And it's no secret that I can't seem to shake the possible all-pervasive role of consciousness. So much, after all, is determined by relative positioning, right down to Einstein's theory of relativity."

"I know it," he says as we again move forward. "Even down to the feature of the multistate—you know, where Nature adapts to choices being made."

He stops, and points to a path that cuts into the woods. "Let's turn in here. I want to show you something."

We turn, head into the winter woods. As the bare tree limbs watch us from above, all sounds suddenly seem to lessen and there is quiet everywhere. "So peaceful," he says.

I nod, and we walk silently along the path for a few minutes, deeper into the woods. "You know," I say finally, my voice naturally lower than before, "both the Copenhagen view *and* the multistate theory hint at a possible heightened role for the observer, and in extension, for consciousness …"

"Explain."

"Let's look at the relative state theory with two assumptions: the first being that the theory itself is indeed true, and the second being that everyone in this r-state is under your study/meditation routine. We know that, again if the theory is true, all our choices that lead to physical actions will experience the branching off. So there will be times that we're angry, and we barely control it, but we know we came so very close to saying or doing something that could have opened a whole can of worms. And if we came too close and our choice of controlling our anger or not controlling it could have easily gone either way, then we will branch off to live out both sides of the possibility. This could clearly lead to some crazy occurrences in some of those universes. In one universe, you assaulted someone and were arrested as opposed to just walking away, which you may have done in this universe. And you can imagine all of the ripple effects of being arrested in that r-state?"

He nods. "Certainly."

"Now, if we live in a certain way in this r-state that is, let's say, thoughtful and tempered, then our branching off, except for the occasional aberration all humans are prone to experience from time to time, like losing our temper, will reflect that as well. If we're gentle and compassionate in one r-state, more times than not we'll be gentle and compassionate to the same degree in the other r-state. These traits will carry over. If we're violent and combative by nature, then these too will carry over. Now, if everyone in this r-state has truly moved into living a more contemplative life, then that contemplative life, along with its greater under-standing and heightened awareness, will reverberate in the other r-states as well. At some point, a real change will begin to occur across the entire multistate."

"And if so," he adds in, "what implications might this hold for the evolution of our species?"

I look at him. "I would think it would accelerate it. If what we said before is true, that Nature may adapt to our new level of understanding, and if this is hap-pening in multiple universes, then …"

"The first question," he holds up a finger, "someone would probably ask at this point would be: there are separate r-states that apparently don't interact with one another, and which we are only aware of one at a time, so how will any of this make a difference in an overall sense?"

I stop in my tracks, turn to him. "Because of the unified field theory."

He nods. "Just what I was thinking. If the unified field theory is true, then these r-states must also be unified at some level."

"And we already know that closely aligned r-states are capable, if this all is true, of occasionally overlapping with one another, as the double-slit experiment seems to indicate. And if we're all evolving in a unified manner with Nature, removing the separation between ourselves and all else, then the r-states would stand a better chance of becoming more closely aligned and thus overlapping."

"And this may be," he concludes with a nod, "where our species' next level of evolution will occur. A passageway into a higher level of understanding, of con-sciousness."

We turn, both nodding silently, and continue walking. We walk for about fif-teen minutes, not a word spoken between us. Suddenly I hear what sounds like a small stream up ahead.

He smiles. "It's been pretty warm the last 24 hours," he says, a bright excite-ment in his tone, "I had a feeling it would be flowing a little."

We move around a small bend in the path and I find myself indeed facing a small, surprisingly quickly flowing stream. He moves ahead of me. "Enough of the snow's melted, so it's flowin' pretty nicely."

"Wow," I pause to look around, to listen, "this is really nice."

"Yes, I come here all the time."

I watch as he removes his gloves, and sits on a huge rock at the edge of the stream. He motions to me. "Have a seat. Enjoy."

I move over to a nearby rock which, because it is dry, has obviously caught a good deal of sunlight through the trees. I ease down, feel my body relax. He smiles over at me. "The little things in life …"

I nod, and we sit in silence for a few minutes.

"Sometimes, I come out here," he says after awhile, "and just close my eyes, and just listen. In the early mornings it's great. The birds are active and their songs fill the forest. Even now you can hear them if you listen."

I close my eyes, let out a deep breath, and hear all of the forest. Nothing is here that does not belong.

"Remove yourself from the equation," I hear him say softly, "and all will be clear."

"Remove yourself from the equation," I whisper inwardly, hearing everything, "and all will be clear."

◆　　　◆　　　◆

Rudy set the packet down, stared over at his group as they sat before the pounding drums. He pulled in a deep breath, closed his eyes and listened …

10

One Week Later

The river flowed behind him. Halfway up the hillside, James paused and turned to look back at the rushing water. He pulled in a deep breath, felt the warm summer air glide into his lungs, and marveled. At that moment, he remembered his father telling him to never lose sight of how many things had to happen just perfectly for him to exist. James smiled, pulled in another breath, and pressed up the hill.

He had decided to take this particular Friday off from work. With almost two weeks of sick days built up, he decided to do something he hadn't done in years—take a Friday off and spend part of the day hiking by himself. As he moved up the hill, his staff pressing into the trail with each and every step, he thought about everything that had been discussed over the last few hikes. There was so much to know, and so much uncertainty, but he was not deterred by this. In fact, he felt enhanced by it. The possibilities that existed were absolutely amazing, and this made the search for what it all meant even more alluring. At one time in his life he would have found both the uncertainty of Nature and some of its more fantastic aspects daunting, but not now. He truly felt, in his blood, that the Truth would someday be revealed through God's Nature.

As he moved closer to the crest of the hill, he thought of a point that had been made in one of his dad's writings. As fantastic and outlandish as some of the theories are that are currently floating around, are they any more inconceivable than some of the beliefs that have guided humankind over the course of the last several thousand years, and that are still guiding us today? Is the idea, for example, of a multiverse any more inconceivable than some of our mainstream ideas regarding an afterlife? Is the idea of a particle traversing every possible path to arrive at a particular destination any more inconceivable than the idea of a soul living through thousands of lifetimes? Is the twin paradox any more inconceivable than the notion of evil spirits and angels guiding the destinies of individuals? Is the idea that the act of observing impacts—and even, to a degree, creates—the reality we experience any more inconceivable than some of our more supernatural ideas of how the world was created? Is the idea of Nature revealing more of itself to our species once a critical mass evolves any more inconceivable than the notion of a supreme being using supernatural powers to reveal himself to the masses? Is the idea of deep observation that reduces the brain's parietal lobe and thus merges one with consciousness any more inconceivable than the voice of God speaking to an individual immersed in prayer?

What's more, the idea that humankind could take to the air in a machine was once considered inconceivable. The idea that our solar system did not constitute the whole of the universe was once considered inconceivable. The idea that the Earth was not flat was once considered inconceivable. The idea that we could land a spacecraft on the moon was once considered inconceivable. The idea that we could identify a criminal two decades later through a tiny piece of genetic material left at the crime scene was once considered inconceivable. And the idea that there could be a worldwide network that could transmit a message from one end of the globe to the other in a matter of seconds using a thing called a computer was once thought to be inconceivable.

And regarding the long periods of uncertainty that the intense study of Nature brings in one's pursuit of Truth, James recalled his father explaining this to him by way of an often used analogy. James could still hear his voice. "Be careful," his father had told him, "that your search for a greater understanding of life and existence doesn't come down to something akin to the story of the man who lost his wallet. One day a man was about to leave for work, but realized just before he walked out that he had forgotten his wallet. In the back of his mind he knew he had probably left the wallet in his bedroom, but he feverishly began to search the living room. He moved the sofa, the coffee table, the television stand, and even a section of the carpet. After watching all of this for several minutes, his wife finally asked him: 'why do you insist on searching through the living room when you're almost certain that you left your wallet in the bedroom?' The man paused and looked up at her—'because the light is brighter in here.' "

James smiled as he finally reached the same rock formation he and Rudy had sat upon during their first Patapsco hike. He pulled his backpack off and eased down onto one of the large flat rocks. He let his body relax while he listened for a long moment to the distant flow of the river.

He sat for a few minutes and took it all in, all the while his mind mulled over the decision he had made. He reached under his shirt, pulled the chain free and let his fingertips trace the etching on the tiny urn. He took a deep breath, held the urn up to his lips and gently kissed it. "I love you, Dad," he whispered, and then placed it back under his shirt.

A moment later, he reached into the front zipper of his backpack and pulled out his phone. His fingers quickly typed out a text message to his brother that consisted of only two words. *"It's time ... "*

He returned the phone to his backpack and then eased his body back until he was lying flat on the smooth rock, staring up at the hazy May sky. He smiled, thinking once again of just how beautiful life was. He took a deep breath, closed his eyes, and let himself drift. He heard the river, the slight breeze, the birds singing in the forest behind him, his own breathing, but then ... he heard something else. Amidst it all, within it all, he could now hear the subtle hum. He relaxed his body further and followed this sound of silence ... He thought of his brother, and how the silence was with him at this very moment, and then he let that knowledge go. He thought of his mother and of Sinclair, and how the silence was with them too, and then he let it go. He thought of his sweet wife and soon to be born child, and knew the silence was with them both, and then he let it go. He heard the hawk now drifting overhead, and knew the silence was guiding his journey, and then let it go. He drifted further until he saw his father's face, his eyes, his smile, and knowing that the silence was there with him, he let him go ...

In memory of Russell J. Cordua, Sr.

Afterword

This author completed and copyrighted an earlier version of this book in 2001. Nearly all of the science-inspired discussions, from the role and impact of the observer/measurer to the multiverse, were included in that 2001 version.

Endnotes

[1]. Wolfe, pgs. 133–139.

[2]. David Deutsch, *The Fabric of Reality* (New York: Penguin Press, 1997), pgs. 40–43.

[3]. Wolfe, p. 139.

[4]. Deutsch, pgs. 43–44.

[5]. Brian Greene, *The Elegant Universe* (New York: Random House, 1999), p. 114.

Bibliography

Amir, Aczel. *Entanglement: The Greatest Mystery in Physics*. New York: Four Walls Eight Windows, 2001.

Arena, Susan, and Morris Hein. *Foundations of College Chemistry*, 7th ed. Pacific Grove, California: Brooks/Cole, 2000.

Asimov, Isaac. *Biographical Encyclopedia of Science & Technology, 2nd ed*. Garden City, New York: Doubleday & Company, 1982.

Austin, James. *Zen and the Brain*. Cambridge, Massachusetts: The MIT Press, 1999.

Barrett, William. *Death of the Soul: From Descartes to the Computer*. New York: Random House, 1986.

Begley, Sharon. "This Year Try Getting Your Brain Into Shape," <u>Wall Street Journal</u>, January 10, 2003.

Begley, Sharon. *Train your Mind, Change Your Brain*. New York: Random House Publishing Group, 2007.

Bennett, Jeffrey, Megan Donahue, Nicholas Schneider, and Mark Voit. *The Essential Cosmic Perspective, 3rd ed.*, San Francisco: Pearson/Addison Wesley, 2005.

Chalmers, David J. *The Conscious Mind: In Search of a Fundamental Theory*. Oxford: Oxford University Press, 1996.

Cole, K.C. *The Hole in the Universe*. New York: Harcourt, Inc., 2001.

Cromie, William J. "Meditation Changes Temperatures: Mind Controls Body in Extreme Experiments," <u>Harvard Gazette</u>, April 18, 2002.

Dennett, Daniel C. *Consciousness Explained*. New York: Little, Brown and Company, 1991.

Deutsch, David. *The Fabric of Reality*. New York: Penguin Press, 1997.

Emoto, Masaru. *The Hidden Messages in Water*. New York: Simon & Schuster, 2005.

Giancoli, Douglas C. *Physics: Principles with Applications, Volume1*. Upper Saddle River, New Jersey: Prentice Hall, 1980.

Goleman, Daniel. *Emotional Intelligence: Why it can matter more than IQ*. New York: Bantam Books, 1995.

Gott, Richard J. *Time Travel in Einstein's Universe: The Physical Possibilities of Travel Through Time*. Boston: Houghton Mifflin Company, 2001.

Greene, Brian. *The Elegant Universe*. New York: Random House, 1999.

Howard, Pierce J. *The Owner's Manuel for the Brain: Everyday Applications from Mind-Brain Research*. Austin, Texas: Bard Press, 2000.

Hawking, Stephen. *A Brief History of Time*. New York: Bantam Books, 1996.

Kane, Gregory. *Supersymmetry: Unveiling the Ultimate Laws of Nature*. Cambridge: Perseus Publishing, 2000.

Lavine, T.Z. *From Socrates to Sartre: The Philosophic Quest*. New York: Bantam Books, 1984.

Magee, Bryan. *The Story of Philosophy*. New York: Dorling Kindersley Publishing, 2001.

Martin Rees. "A Field Guide to the Invisible Universe," Discover Dec. 2003: 42.

Post, Richard. *Chemistry: Concepts & Problems*. New York: John Wiley & Sons, 1996.

Seeds, Michael A. *Astronomy: The Solar System and Beyond*. Pacific Grove, California: Brooks/Cole, 2001.

Simpkins, Alexander C., and Annellen M. Simpkins. *Living Meditation: From Principle to Practice*. Boston: Charles E. Tuttle Co., 1997.

Smolin, Lee. *Three Roads to Quantum Gravity*. New York: Basic Books, 2001.

Stenger, Victor. *Timeless Reality*. New York: Prometheus Books, 2000.

Tipler, Paul. *Physics*. New York: Worth Publishers, 1976.

Watts, Alan. *Still the Mind: An Introduction to Meditation*. Novata, California: New World Library, 2000.

Ward, Mark. *Beyond Chaos: The Underlying Theory Behind Life, the Universe, and Everything*. New York: St. Martin's Press, 2001.

Wheeler, John Archibald (with Kenneth Ford). *Geons, Black Holes & Quantum Foam: A Life in Physics*. New York: W.W. Norton & Company, 1998.

Wolfson, Richard. *Einstein's Relativity and the Quantum Revolution: Modern Physics for Non-Scientists*. Chantilly, Virginia: The Teaching Company, 2000.

Wolf, Alan. *Taking the Quantum Leap: The New Physics for Non-Scientists*. New York: Harper & Row, 1989.

978-0-595-46313-8
0-595-46313-4